CET Certified in Emerging Technology

AF817957

BLOCKCHAIN
FUNDAMENTALS
Study Guide

Part of the CET Stackable Certificate Program
www.isaca.org/cet

ISACA®

About ISACA

For more than 50 years, ISACA® (*www.isaca.org*) has advanced the best talent, expertise and learning in technology. ISACA equips individuals with knowledge, credentials, education and community to progress their careers and transform their organizations, and enables enterprises to train and build quality teams. Among those credentials, ISACA advances and validates business-critical skills and knowledge through the globally respected Certified Information Systems Auditor® (CISA®), Certified in Risk and Information Systems Control® (CRISC®), Certified Information Security Manager® (CISM®), Certified in the Governance of Enterprise IT® (CGEIT®) and Certified Data Privacy Solutions Engineer™ (CDPSE™) credentials. ISACA is a global professional association and learning organization that leverages the expertise of its 145,000 members who work in information security, governance, assurance, risk and privacy to drive innovation through technology. It has a presence in 188 countries, including more than 220 chapters worldwide.

Disclaimer

ISACA has designed and created *Blockchain Fundamentals Study Guide* (the "Work") primarily as an educational resource for professionals. ISACA makes no claim that use of any of the Work will assure a successful outcome. The Work should not be considered inclusive of all proper information, procedures and tests or exclusive of other information, procedures and tests that are reasonably directed to obtaining the same results. In determining the propriety of any specific information, procedure or test, professionals should apply their own professional judgment to the specific circumstances presented by the particular systems or information technology environment.

Reservation of Rights

© 2021 ISACA. All rights reserved. No part of this publication may be used, copied, reproduced, modified, distributed, displayed, stored in a retrieval system or transmitted in any form by any means (electronic, mechanical, photocopying, recording or otherwise) without the prior written authorization of ISACA.

ISACA
1700 E. Golf Road, Suite 400
Schaumburg, IL 60173, USA
Phone: +1.847.660.5505
Fax: +1.847.253.1755
Contact us: https://support.isaca.org
Website: www.isaca.org

Provide Feedback: To comment, please visit *https://support.isaca.org.*
Participate in the ISACA Online Forums: https://engage.isaca.org/onlineforums

Twitter: http://twitter.com/ISACANews
LinkedIn: www.linkedin.com/company/isaca
Facebook: www.facebook.com/ISACAGlobal
Instagram: www.instagram.com/isacanews/

ISBN 978-1-60420-757-6
Blockchain Fundamentals Study Guide
Printed in the United States of America

Acknowledgments

The *Blockchain Fundamentals Study Guide* is the result of the collective efforts of many volunteers. Special thanks go to:

Lead Developer

Ron Quaranta, EC–Information Systems and Technology Project Management, Wall Street Blockchain Alliance, USA

Expert Reviewers

Marianne Azer, Ph.D., Member of the Egyptian Parliament, Egypt
Vladlena Benson, Ph.D., CDPSE, United Kingdom
Gerard (Rod) Brennan, Ph.D., CFE, Lukka, USA
Earnest Collins, CISA, CISM, CFE, CISSP, AES, USA
Ramsés Gallego, CISM, CGEIT, CCSK, CISSP, Six Sigma Black Belt, Micro Focus, Spain
Sonia Goklani, CEO, CLEARTRACK, USA
Niel Harper, CISA, CRISC, CDPSE, CISSP, Denmark
Will Janensch, USA
Luis Jugo, CISA, CCSA, CIA, CRMA, PMP, USA
Neil Lappage, CISM, CISSP, Australia
Tuan Phan, CISSP, CBSP, Security+, PMP, Caplock Security, USA
Marc T. Staut, CITO, Boomer Consulting, USA

Board of Directors

Tracey Dedrick, Chair, Former Chief Risk Officer, Hudson City Bancorp, USA
Rolf von Roessing, Vice-Chair, CISA, CISM, CGEIT, CDPSE, CISSP, FBCI, Partner, FORFA Consulting AG, Switzerland
Gabriela Hernandez-Cardoso, Independent Board Member, Mexico
Pam Nigro, CISA, CRISC, CGEIT, CRMA, Vice President–Information Technology, Security Officer, Home Access Health, USA
Maureen O'Connell, Board Chair, Acacia Research (NASDAQ), Former Chief Financial Officer and Chief Administration Officer, Scholastic, Inc., USA
David Samuelson, Chief Executive Officer, ISACA, USA
Gerrard Schmid, President and Chief Executive Officer, Diebold Nixdorf, USA
Gregory Touhill, CISM, CISSP, President, AppGate Federal Group, USA
Asaf Weisberg, CISA, CRISC, CISM, CGEIT, Chief Executive Officer, introSight Ltd., Israel
Anna Yip, Chief Executive Officer, SmarTone Telecommunications Limited, Hong Kong
Brennan P. Baybeck, CISA, CRISC, CISM, CISSP, ISACA Board Chair, 2019-2020, Vice President and Chief Information Security Officer for Customer Services, Oracle Corporation, USA
Rob Clyde, CISM, ISACA Board Chair, 2018-2019, Independent Director, Titus, and Executive Chair, White Cloud Security, USA
Chris K. Dimitriadis, Ph.D., CISA, CRISC, CISM, ISACA Board Chair, 2015-2017, Group Chief Executive Officer, INTRALOT, Greece

ISACA. All Rights Reserved.

Page intentionally left blank

ISACA. All Rights Reserved.

Table of Contents

List of Figures 9

An Introduction to Blockchain 11

Chapter 1:
Blockchain Principles 13

Blockchain Principles 14
- **1.1 Learning Objectives** 14
- **1.2 Blockchain Beginnings** 14
- **1.3 Blockchain Defined** 15
- **1.4 Distributed Ledger vs. Blockchain** 15
- **1.5 Blockchain Processes** 15
  - 1.5.1 Encryption 16
  - 1.5.2 Trust 18
  - 1.5.3 Transactions 18
  - 1.5.4 Mining 18
  - 1.5.5 Forking 19
  - 1.5.6 Block Addition 19
  - 1.5.7 Consensus Methodology 20
  - 1.5.8 Consensus Algorithms 21
    - Proof of Work 21
    - Proof of Stake 21
    - Practical Byzantine Fault Tolerance (pBFT) 21
    - Proof of Elapsed Time 21
    - Proof of Importance (PoI) 22
  - 1.5.9 Determining Consensus Needs 22
  - 1.5.10 Cryptographic Protocols 22
- **1.6 Chapter 1 Knowledge Check** 24

Chapter 2:
Cryptocurrency 29

Cryptocurrency 30
- **2.1 Learning Objectives** 30
- **2.2 Overview** 30
- **2.3 Types of Virtual Currencies** 31
  - 2.3.1 Altcoins 32
  - 2.3.2 Cryptoassets 33
    - Security Tokens 33
    - Utility Tokens 33
    - Stablecoins 34
  - 2.3.3 Cryptocurrency Benefits 34
  - 2.3.4 Smart Contracts 34
- **2.4 Chapter 2 Knowledge Check** 38

Chapter 3:
Topology and Blockchain Models 41

Topology and Blockchain Models 42
- **3.1 Learning Objectives** 42
- **3.2 Overview** 42

ISACA. All Rights Reserved.

3.3 **Public Blockchains** 42
3.4 **Private Blockchains** 43
3.5 **Hybrid Blockchains** 44
3.6 **Model Comparisons** 45
3.7 **Chapter 3 Knowledge Check** 46

Chapter 4:
Blockchain IT Infrastructure 49

Blockchain IT Infrastructure 50
4.1 **Learning Objectives** 50
4.2 **Overview** 50
4.3 **Blockchain Infrastructure** 50
4.3.1 Blockchain Nodes 54
Full nodes 55
Miners 55
Listening nodes 55
Masternodes 55
4.3.2 Blockchain Network Interoperability 55
Data Exchange 56
Digital Asset Exchange 56
4.3.3 Blockchain Specific Hardware 59
4.4 **Chapter 4 Knowledge Check** 61

Chapter 5:
Business Usage and Considerations 65

Business Usage and Considerations 66
5.1 **Learning Objectives** 66
5.2 **Overview** 66
5.3 **Major Blockchain Platforms** 66
5.3.1 Amazon Web Services (AWS) 66
5.3.2 Ethereum 66
5.3.3 Hyperledger 67
5.3.4 Microsoft Azure 67
5.3.5 R3 67
5.4 **Example Use Cases** 67
5.4.1 Financial Markets 67
Clearing and Settlement 68
5.4.2 Supply Chain 68
Walmart and the IBM Food Trust 68
5.5 **Healthcare** 69
5.5.1 BurstIQ and Patient Records 69
5.6 **Other Industries** 70
5.7 **Tokenization and Digital Assets** 70
5.8 **Token Technology Standards** 71
5.8.1 Native Tokens 71
5.8.2 Layer 2 Tokens 71
ERC-20 71
ERC-721 71
5.8.3 Token Types 71
Central Bank Digital Currency (CBDC) 71
Cryptocurrency 72
Security Token 72
Stablecoins 72
Tokenized Asset 72

ISACA. All Rights Reserved.

Utility Tokens 72
5.8.4 Benefits of Tokenization 72
5.8.5 Tokenization Example: Real Estate 73
5.9 Blockchain Validity and Audit Considerations 73
5.10 Conclusion 74
5.11 Chapter 5 Knowledge Check 75
Glossary 79

ISACA. All Rights Reserved.

Page intentionally left blank

ISACA. All Rights Reserved.

LIST OF FIGURES

Chapter 1. Blockchain Principles

Figure 1.1—Public/Private-Key Asymmetric Cryptography ... 16

Figure 1.2—Changing a Public Key to a Bitcoin Address ... 17

Figure 1.3—How a Bitcoin Transaction Works ... 20

Figure 1.4—Blockchain Terminology ... 23

Chapter 2. Cryptocurrency

Figure 2.1—Applying Business Logic With Smart Contracts ... 36

Chapter 4. Blockchain IT Infrastructure

Figure 4.1—Blockchain Infrastructures ... 50

Figure 4.2—Client Server and Peer-to-Peer Network Architecture ... 51

Figure 4.3—Linked Lists ... 52

Figure 4.4—Pointer/Linked List Paradigm ... 53

Figure 4.5—Components of a Block ... 53

Figure 4.6—Bitcoin Block Number Example ... 54

Figure 4.7—Ethereum Block Example ... 54

Figure 4.8—Atomic Swaps ... 57

Figure 4.9—Computational Power for Processing Speed and Power ... 60

ISACA. All Rights Reserved.

Page intentionally left blank

ISACA. All Rights Reserved.

An Introduction to Blockchain

Blockchain Fundamentals Study Guide presents a beginning study of blockchain technology, developed with the information technology professional in mind. Although it is not meant to be a deeply technical reference or programming guide, this study guide provides the reader with a broad and deep understanding of blockchain technology—its principles, terminology and history, and current enterprise practices, i.e., in the real world.

This study guide briefly presents the origins of blockchain—its original and current goals and the importance of its structure and implementation. The reader will learn about the multitude of blockchain solutions now globally available, how they interact or interoperate, and how they may evolve in the future. This study guide also reviews and analyzes the types of consensus methodologies, the importance of determining enterprise consensus needs, and blockchain validity and audit considerations. Finally, no study of the blockchain realm is complete without a discussion of cryptoassets (including cryptocurrencies) and tokenization, which promises not only to progress how assets are exchanged between parties but also to potentially enable previously undiscovered business models. Each chapter ends with a knowledge-check section.

At the completion of this study guide, learners, will have a fundamental knowledge of blockchain technology and the technology it inspired, such as cryptocurrency, and the basic tools to use this innovation in their enterprises and their careers.

The *Blockchain Fundamentals Study Guide* helps individuals prepare for the ISACA® Fundamentals certificate exam, one of the components of the ISACA Certified in Emerging Technology™ certification program.

Every chapter contains the following:

- **Learning objectives**—Knowledge that learners should understand after reading the chapter
- **Sections**—Knowledge content
- **Knowledge checks**—Quiz on the chapter knowledge

ISACA. All Rights Reserved.

Page intentionally left blank

ISACA. All Rights Reserved.

BLOCKCHAIN
FUNDAMENTALS
An ISACA Certificate

Chapter 1:

Blockchain Principles

Blockchain Principles

1.1 Learning Objectives 14
1.2 Blockchain Beginnings 14
1.3 Blockchain Defined 15
1.4 Distributed Ledger vs. Blockchain 15
1.5 Blockchain Processes 15
1.6 Chapter 1 Knowledge Check 24

ISACA. All Rights Reserved.

Blockchain Principles

1.1 Learning Objectives

After completing this chapter, learners will be able to:

- Explain the history of blockchain technology.
- Describe the evolution of blockchain and cryptocurrencies as we know them today.
- Articulate differences between distributed ledger technology and blockchain.
- Describe processes that are used to establish and maintain a blockchain.
- Explain encryption techniques that are used in a blockchain system.
- Identify different consensus methodologies that blockchain systems use.

1.2 Blockchain Beginnings

To be able to understand and discuss blockchain and the technological principles behind it, one must understand money or at least electronic money in the modern world. Indeed, the first discussions about electronic peer-to-peer money came before the large-scale implementation of blockchain technology that is now recognized as the Bitcoin cryptocurrency system. Several versions of cryptocurrency were in use before bitcoin.

Note: In this study guide, the capitalized term Bitcoin refers to the underlying protocol and blockchain; the lowercase term bitcoin denotes the cryptocurrency.

The cryptographer David Chaum[1] produced extensive work on the development of a true electronic cash, based on his concerns regarding personal privacy and money. These concerns evolved into his theory that the world needs a tokenized version of cash that is as private and secure as fiat money. To accomplish this, he developed methods of data encryption, many of which continue to be used for the Internet today (e.g., portions of the RSA algorithm). The outcome of this extensive work resulted in the creation of DigiCash, an electronic money product and corporation with a mission to deliver private, secure digital money, but the corporation declared bankruptcy and ceased operations.[2, 3]

The advent of DigiCash kicked off a wave of new startups that sought to enter the digital money space throughout the 1990s. This coincided with the Dot.com boom that brought vast amounts of venture money to a variety of Internet-based business models, including E-Gold,[4] which ran into significant legal trouble, and PayPal.[5]

In 2008, an anonymous technologist and author (or authors) under the pseudonym of Satoshi Nakamoto published, on a technology mailing list, a paper that calls for the development of a cryptographically enabled currency that is independent of national monetary systems.[6]

In January 2009, the first version of the open-source Bitcoin software, was made publicly available. The Bitcoin blockchain launch initiated the first broad use of a cryptographic, peer-to-peer system to allow for the secure and anonymous exchange of value between parties. Its launch has created a more than decade-long discussion about the

1 Wikipedia, "David Chaum," *https://en.wikipedia.org/wiki/David_Chaum*
2 Pitta, J.; "Requiem for a Bright Idea," *Forbes*, 1 November 1999, *www.forbes.com/forbes/1999/1101/6411390a.html*
3 Microsoft offered to pay over $180 million to include DigiCash in every Windows PC in 1998. Chaum declined, and the DigiCash company declared bankruptcy and ceased operations not long afterwards.
4 Zetter, K.; "Bullion and Bandits: The Improbable Rise and Fall of E-Gold," *Wired*, 9 June 2009, *www.wired.com/2009/06/e-gold/*
5 Funding Universe, "PayPal, Inc., History," *www.fundinguniverse.com/company-histories/paypal-inc-history/*
6 Nakamoto, S.; "Bitcoin: A Peer-to-Peer Electronic Cash System," Bitcoin.org, *https://bitcoin.org/bitcoin.pdf*

ISACA. All Rights Reserved.

nature of trust, money and identity, and a new wave of innovative startups looking to reinvent whole economic models based on blockchain technology.

1.3 Blockchain Defined

Blockchain is a shared transactions ledger that can be accessed by and among multiple parties, using cryptography and peer-to-peer technology to secure data into blocks and store them in an immutable chain of transactions, without any trusted central authority. By design, each participant in the blockchain possesses the same copy of this ledger and has the ability to write to, or modify this ledger, assuming that they are properly authorized to do so. The fundamental function of blockchain is to accomplish this sharing and distribution of data or value, without the need for a trusted intermediary, and without any enforced system management. It is the design of the software and network itself that allows participants to trust the accuracy and veracity of the information on the blockchain. This technology and its benefits are examined in the following sections.

1.4 Distributed Ledger vs. Blockchain

Throughout history, people have been using some form of ledger to record various types of transactions, for example, ledgers that are used to account for financial or accounting transactions. The ledgers require third parties who can attest to the accuracy and completeness of records. In this context, ledgers have been in use for centuries.

Distributed ledgers are ledgers that exist across multiple participants or locations. Distributing the ledger data across multiple participants eliminates the need for a central authority or intermediaries to process, validate or authenticate transactions. In contrast, many enterprises continue to use centralized databases that represent single vectors of risk and points of failure. Recent major data hacks are examples of the challenges of centralized databases in the modern world.[7] Distributed ledger technology removes single vectors of risk and points of failure.

The blockchain distributed ledger differs from a typical distributed ledger because the blockchain uses a chain of data blocks to maintain that data and provide valid, secure consensus.

Thus, blockchain technology is considered a form, or subset, of distributed ledger technology. Like distributed ledgers, a blockchain is distributed across nodes and managed within peer-to-peer networks. Being a form of distributed ledger, it can exist without the management of a centralized authority, and the data sets and data quality are preserved via database replication and computational trust. It is the structure of a blockchain that makes it distinct from other types of distributed ledgers. On a blockchain, data is assembled together and organized into blocks, which are then linked to one another sequentially and secured using cryptography. A blockchain is a continuously growing chain of records. An append-only rule allows data to be added to this ledger database. Cryptography secures the addition of data; therefore, altering or deleting previously entered data in earlier blocks is impossible. Because of these features, blockchain technology has been considered a new platform for everything from managing data records, to recording events, to processing asset transfers and more.

1.5 Blockchain Processes

This section describes the detailed mechanisms for adding transactions to a blockchain, by using the Bitcoin blockchain.[8]

The Bitcoin blockchain is a shared public ledger, and the entirety of the Bitcoin network is dependent on this public ledger. All transactions that are confirmed by the network are included in the blockchain.

[7] Leonhardt, M.; "The 10 biggest data hacks of the decade," CNBC, 27 December 2019, *www.cnbc.com/2019/12/23/the-10-biggest-data-hacks-of-the-decade.html*

[8] *Ibid.*

ISACA. All Rights Reserved.

1.5.1 Encryption

Bitcoin, and indeed many public blockchains, make use of public/private key encryption to help manage ownership and transactions on the blockchain.

Public/private key encryption leverages asymmetric cryptography using a public key and a private key to accomplish different tasks, such as sending or receiving bitcoin. Public keys can be widely shared; private keys must be kept secret. However, it is critical to note that a private key can never be reset or recovered if lost or stolen. (This differs from symmetric cryptography, in which a key can be shared between sender and receiver). **Figure 1.1** provides a basic representation of public/private-key asymmetric cryptography.

Figure 1.1—Public/Private-Key Asymmetric Cryptography

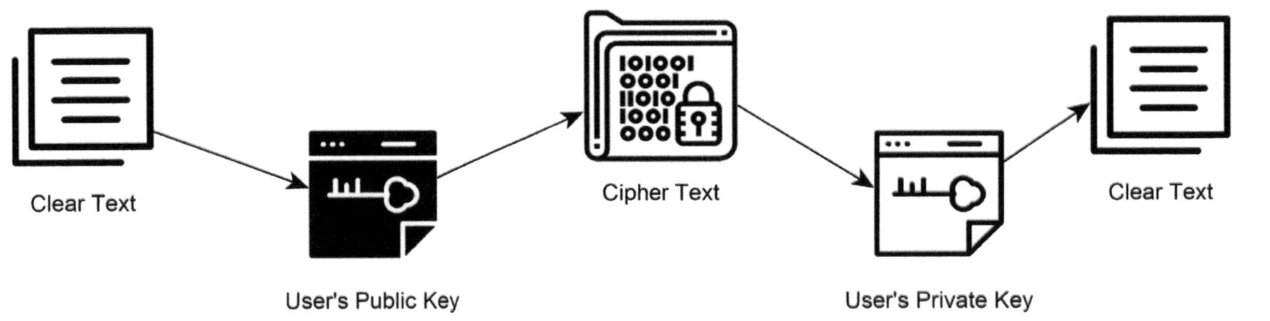

A message is encrypted using a receiver's public key so that only the receiver with the related private key can decrypt and read it. Conversely, using the sender's private key, a digital signature is created so that the recipient of a digitally signed message can verify, by using the sender's public key, that the message was created by the owner of the private key, i.e., the sender.

Each bitcoin is associated with its current owner's public key. When someone sends bitcoin, a transaction is created connecting the new owner's public key to the specific amount of bitcoin, and it is signed (not shared) with the sender's private key. When this transaction is broadcast to the Bitcoin network, the network is aware that the new owner of these coins is the owner of the new public key, i.e., the recipient.

Most cryptocurrencies use a wallet to store the public/private key pairings that allow a user to track ownership, and send or receive cryptocurrencies. A wallet can be a hardware device (e.g., a Ledger Nano[9]), a program or a service. Cryptocurrency itself does not reside in any wallet, because cryptocurrencies do not exist in a physical form. Rather, cryptocurrencies are maintained on the blockchain, which consists of transaction records that detail the private and the public keys that have control over these cryptocurrencies.

A Bitcoin wallet address and the public key are not the same, but they are mathematically related. A Bitcoin wallet address is a hashed version of a public key. A hash is a function that converts an input of letters and numbers, the public key in this case, into an encrypted output of fixed length. Hashing is a critical function of blockchain usage.

Bitcoin transactions are intended to be anonymous. This anonymity is provided by using a permutation of the user's public key as the user's identifier (user address) and the recommendation that the sender in the transaction employ a different public/private key pair for each transaction. If the same address is used for repeated transactions, it might be possible to determine the user's identity through observable transaction patterns. For anonymity and ease of use, the user's native public key is hashed and converted into a unique string that is shorter; contains alphabetic letters and numbers, instead of only binary digits; and does not include difficult-to-differentiate letters and numbers.[10]

[9] Ledger, *www.ledger.com*
[10] ISACA, *Blockchain Fundamentals: An Inside Look at the Technology With the Potential to Impact Everything*, USA, 2017, *https://www.isaca.org/bookstore/internet-and-related-security/wbcb*

ISACA. All Rights Reserved.

Figure 1.2 shows the process for transforming a user's public key to a Bitcoin blockchain wallet address.

Figure 1.2—Changing a Public Key to a Bitcoin Address

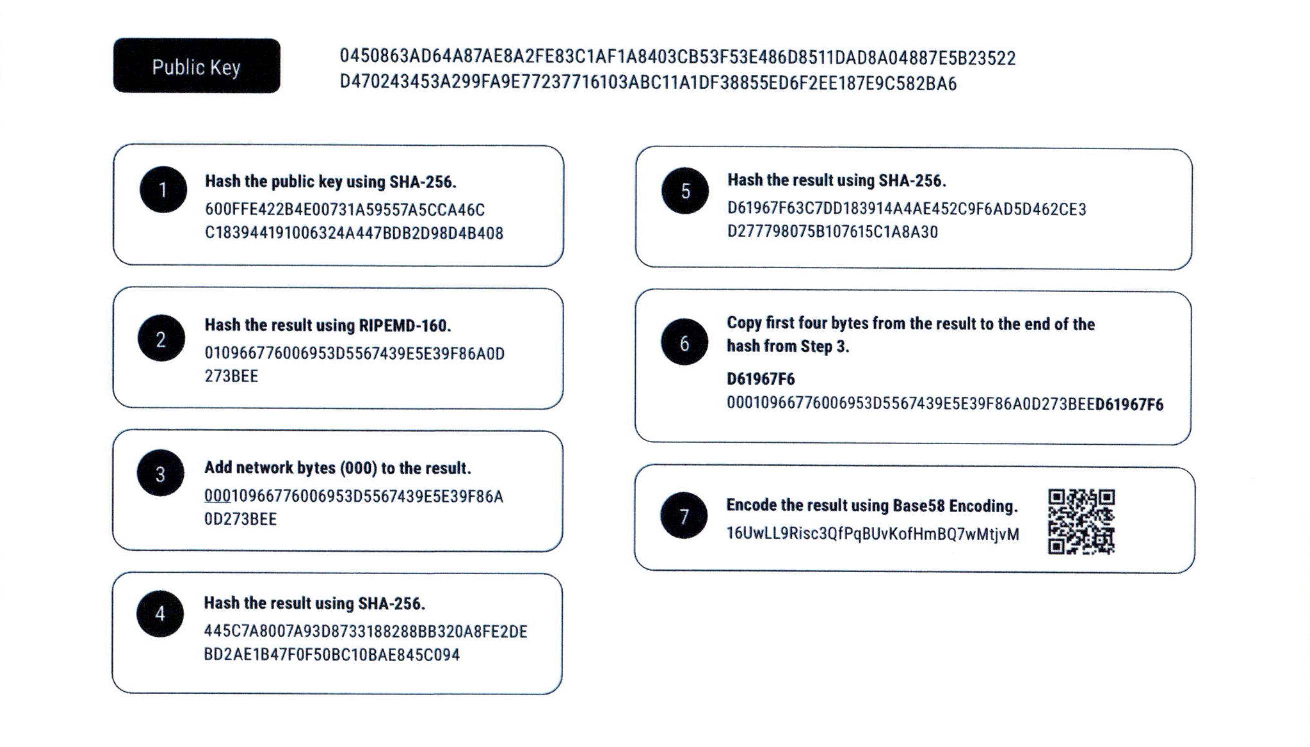

Source: ISACA, *Blockchain Fundamentals: An Inside Look at the Technology With the Potential to Impact Everything*, USA, 2017, figure 4, *https://www.isaca.org/bookstore/internet-and-related-security/wbcb*

Because a blockchain can only contain validated transactions (to prevent fraudulent transactions and double spending of cryptocurrency), it requires the data in a block to be processed through a mathematical hash function, which results in an encrypted output of fixed length. The encrypted output significantly increases the security of transactions, because anyone trying to illicitly decrypt the hash (and hence alter the transaction) is not able to determine the original input.

A wallet address is like a type of account number. There is no harm in giving someone else this account number, if they perhaps want to send someone bitcoin. In the cryptocurrency world, if somebody wants to send another party cryptocurrency, they simply need the relevant wallet address. Because no two wallets are the same, there is generally no chance that a wrong party will receive the cryptocurrency. There is no limit to the number of wallet addresses that can be created. Following is an example of a Bitcoin wallet address, which is believed to be a wallet address belonging to Bitcoin's creator, Satoshi Nakamoto:[11]

1A1zP1eP5QGefi2DMPTfTL5SLmv7DivfNa

A wallet address uses a combination of letters and numbers, both upper case and lower case (created by hashing the user's public key). Since most blockchains are transparent, it is relatively easy to determine the quantity of a given cryptocurrency that a wallet records and previously made transactions. The cryptocurrency wallet address does not reveal the real-world identity of an owner. This is why public blockchains are often called pseudonymous.

[11] Sedgwick, K.; "People Keep Sending Satoshi Nakamoto Bitcoin," 24 December 2017, *https://news.bitcoin.com/people-keep-sending-satoshi-nakamoto-bitcoin/*

ISACA. All Rights Reserved.

1.5.2 Trust

In a blockchain system, the trust needed to maintain the network is embedded within the architecture of the network itself, rather than with a centralized system or third party, which has inherent risk (e.g., large scale hacks like Yahoo[12]) or represents a single point of failure. With blockchain, the participants are not trusted by default, particularly in a permissionless design. Thus, trust in the blockchain stems from each participant adhering to the rules of the network, with all other participants helping to police enforcement among each other. Three distinct functions are associated with the participants in the blockchain:

- **Validation**—Identifying the existence of a transaction and the parties' participation in the network
- **Transactions processing**—Participating in the processing and addition of data to the blockchain while preventing duplicate transactions (also known as double spending)
- **Recording**—Containment of the record of transactions, parties involved and other relevant transaction details

1.5.3 Transactions

Transactions are initiated when sent to a node on the blockchain network. A node possesses the instruction set for how a transaction must be validated. All nodes for transaction initiation are equivalent; therefore, it makes no difference which node initiates or receives a transaction. Transactions not meeting required blockchain criteria are discarded. If not discarded, the transaction is signed with the initiator's private key and validated so that it can be added to the network. An example transaction is the transfer of a quantity of bitcoin from one wallet address to another.

The transaction is then broadcast to other nodes on the peer-to-peer network, and each node will work to validate the transaction. The intrinsic protocols of the blockchain system necessitate that the transaction is dispersed into the network and efficiently reaches all nodes in a relatively short time (dependent on the network). There is no inherent or implied trust between nodes (except in private blockchains, which are discussed in chapter 4).

1.5.4 Mining

The role of miners is important to the blockchain. Mining is the process by which transactions are added to a public ledger, such as the Bitcoin blockchain. Mining is intended to be a computationally exhaustive part of the proof of work consensus process (which is defined later) and allows for the number of blocks to be correctly added to the blockchain in a way that is consistent and ordered. Mining enables the history of all transactions on a blockchain to be added in a way that makes it impossible to modify by anyone.

Given that mining is computational expensive, the reward of bitcoins incentivizes miners to monitor and maintain the blockchain. In this way, network nodes (miners) introduce new bitcoin to the ecosystem, akin to the way a central bank would print money, by upward adjusting bank balances with the US Federal Reserve Bank, for example. (Though bitcoin purists argue that the comparison is rather thin and ends there). Miners can also be paid transaction fees, as a way of subsidizing them for newly created bitcoins that are disseminated to the ecosystem in a decentralized way. This study guide addresses consensus methodologies (of which proof of work, the consensus methodology of the Bitcoin blockchain, is but one of many) and the types of incentivization, later in this study guide. Miners play a critical role in maintaining the security and ongoing viability of a public blockchain.

Each block in a blockchain contains a collection of transactions that have been sent since the previous block. To preserve the integrity of the blockchain, each block in the chain ratifies the veracity of the previous one, all the way back to the first block, usually known as the genesis block. In this way, no party can overwrite preceding records in

[12] Perlroth, N.; "Yahoo Says Hackers Stole Data on 500 Million Users in 2014," *New York Times,* 22 September 2016, *www.nytimes.com/2016/09/23/technology/yahoo-hackers.html*

ISACA. All Rights Reserved.

blocks. When the block (header, transaction and its payload data) is written to a sufficient population of nodes, again determined by the consensus protocol, it will eventually persist in every copy of the blockchain on every node in the network. In principle, every node must add the new block to the end of its blockchain.

1.5.5 Forking

There can be instances when a node no longer adds a block to the end of a particular blockchain, which is referred to as forking. There are two types of forking—a hard fork and a soft fork.

A hard fork is a drastic change to a blockchain network protocol that renders blocks and transactions that were previously considered valid as invalid (or vice versa). A hard fork requires all nodes or users to upgrade to the latest version of the protocol software. This creates a permanent deviation from the previous version of the blockchain. Adding a new rule to the code effectively creates a fork in the blockchain: one path follows the new, upgraded blockchain, and the other path continues along the old path. For the most part, the old chain participants recognize that the version of the blockchain with which they are interacting is either irrelevant or outdated, and will upgrade to the newer version. Alternatively, participants may continue supporting the old blockchain, resulting in the creation of a new blockchain network, like with the case of Ethereum and Ethereum Classic blockchains over the DAO hack.[13]

A soft fork represents a modification to software protocol in which only prior valid blocks and transactions are made invalid. Because old nodes will recognize the new blocks as valid, a soft fork is backward-compatible. A soft fork requires only a majority of the miners to upgrade so that the new rules can be enforced; whereas, a hard fork requires that all nodes upgrade and agree on the new version. Soft forks have been used on the Bitcoin and Ethereum blockchains, among others, to add new and upgraded functions that are backward compatible.

1.5.6 Block Addition

Pending transactions that are ultimately added to a blockchain are confirmed via the mining process. Mining enforces the chronological order in the blockchain and allows different computers to agree on the state of the data in the system, while protecting the neutrality of the network. To achieve a confirmed status, a transaction must be added into a block that fits strict cryptographic rules that will be verified by the network. Such rules prevent previous blocks from being modified, because doing so invalidates all subsequent blocks. Mining also generates the equivalent of a competitive sweepstakes that prevents any one individual from adding new blocks consecutively to the blockchain. In this way, no group or individual can control what is included in the blockchain or substitute parts of the blockchain to reflect changed quantities in their favor, for example.

By design, adding a block to the blockchain is difficult, requiring processing power and time to achieve. To incentivize this work, the Bitcoin blockchain rewards the party, i.e., miner, who manages to produce a block that is added to the blockchain. This reward is in two parts. The first part is a prize of a predetermined number of bitcoins (determined at network design and launch) that is given to the producer of the block, i.e., miner. As of this writing, this prize is now 6.25 bitcoins, and this amount decreases over time at a preset schedule. The second part of the reward is transaction fees. If any transaction fees are a part of the transactions included in the affirmed block, they can be claimed by the block producer. The existing Bitcoin network rules are designed so that the difficulty is adjusted to keep block production to approximately one block per 10 minutes. Thus, the more miners who participate in this mining activity, the more difficult it becomes for each individual miner to produce a block. The higher the total difficulty, the more difficult it also becomes for an attacker to change the blockchain data. Mining is also the device by which bitcoins get created and distributed among stakeholders in the bitcoin ecosystem. Bitcoin was designed so that a total of only 21 million bitcoins will be ever created.

Figure 1.3 shows the multiple steps of the blockchain processes.

[13] Wikipedia, "Ethereum Classic," *https://en.wikipedia.org/wiki/Ethereum_Classic*

ISACA. All Rights Reserved.

Figure 1.3—How a Bitcoin Transaction Works

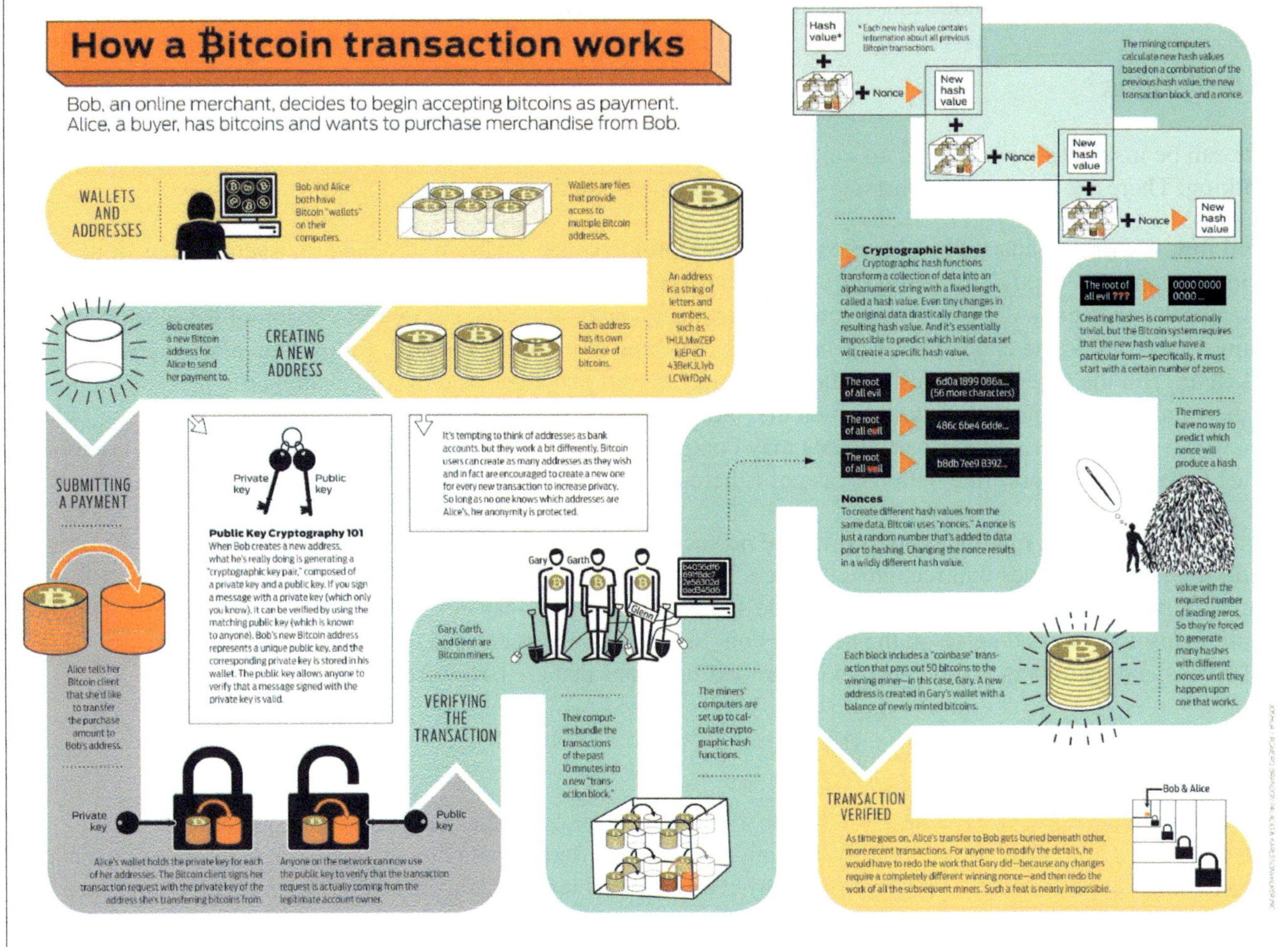

Source: Romero, J.J.; "This Is How a Bitcoin Transaction Works," Brandon Palacio & Karlssonwilker Inc., IEEE Spectrum, 31 May 2012, *https://spectrum.ieee.org/static/how-a-bitcoin-transaction-works*

1.5.7 Consensus Methodology

Although the origins of consensus begin in the social sciences, the focus in this study guide is on its application to computer science. The problem of consensus appears when attempting to control and keep multiple actors in a network in synchronization, whereby resilience and fault tolerance are a requirement for the system or network to operate. Consensus methodology in the realm of computer science got its start with a 1982 paper, titled "The Byzantine Generals' Problem."[14] The term Byzantine refers to the Byzantine Empire, from approximately the fourth through the fifteenth centuries, in which popular historical reference suffered from a number of conspiracies and intrigues among its rulers.

The Byzantine generals' problem in the context of computer science requires us to consider how the majority of the parties (or nodes) within a distributed network can agree on specific data and accomplish the same task to maintain the network. Even more challenging is the need to guarantee that no party or parties in the network can act maliciously to cause a failure of the network.

[14] Lamport, L.; R. Shostak; M. Pease; "The Byzantine Generals Problem," November 1981, *https://lamport.azurewebsites.net/pubs/byz.pdf*

ISACA. All Rights Reserved.

For this reason, blockchain (and all distributed ledger technology and other technologies) seek to be Byzantine Fault Tolerant (BFT). BFT is the property of a system such that the system is able to withstand any failures and can continue to function even if some nodes fail or act maliciously. There are several possible solutions to the Byzantine generals' problem and therefore multiple ways of achieving a BFT-enabled system. The most prominent way that blockchain technology achieves this is via consensus algorithms, the most prominent of which are reviewed in the following sections.

1.5.8 Consensus Algorithms

The foundations of the consensus methodology and its applicability to blockchain has given rise to a multitude of different consensus methods or algorithms, some of which were referenced earlier in this guide. These algorithms define how a blockchain operates, how transactions and data are persisted into the blockchain and how the rules for governance on the blockchain network are enforced.

Proof of Work

Proof of work (PoW) is conducted through miners (participants that keep the blockchain running by providing computing resources) who are competing to solve a cryptographic problem (hash puzzle). This means that the PoW algorithm is used to confirm transactions and produce new blocks that are added to the chain. With PoW, miners compete against each other to complete transactions on the network and get rewarded. The computational work required to accomplish this is fairly (and usually increasingly) difficult for miners to perform but easy for the network to verify. As difficulty increases over time, the amount of computational power, and hence energy consumption, grows. Bitcoin is the first widespread application use of PoW. PoW is applicable to public blockchains.

Proof of Stake

Proof of stake (PoS) is a type of consensus algorithm by which a cryptocurrency blockchain network aims to achieve distributed consensus. In PoS consensus, the creator of the next block of data is chosen via several combinations of random selection and wealth or age (i.e., the stake) within the blockchain. Miners can mine or validate block transactions based on the amount of cryptocurrency a miner holds. PoS was created as an alternative to PoW, which requires large amounts of energy. PoS gives mining power based on the percentage of cryptocurrency held by a miner, which has less risk in terms of network attacks and security. PoS is used only for public blockchains.

Practical Byzantine Fault Tolerance (pBFT)

Practical Byzantine fault tolerance (pBFT) is a consensus mechanism in which all nodes are ordered in sequence with one node being primary node or leader, and all others referred to as backup nodes. All nodes in pBFT systems communicate with one another with the goal being that all honest nodes will come to an agreement of the state of the system using a majority rule. Nodes communicate for two reasons: to prove that messages came from a specific peer node and to confirm that a message was not modified during transmission. pBFT can be used for private and public blockchains and allows for instant transaction finality; however, this methodology requires a great number of messages between nodes, which makes a large blockchain network challenging.

Proof of Elapsed Time

Proof of elapsed time (PoET) is a consensus mechanism algorithm that is often used on the permissioned blockchain networks to randomly decide the next block publisher. The selection process works as follows:

- Each publisher waits for a random amount of time.
- The first publisher to finish the waiting time gets to be the leader for the new block.

ISACA. All Rights Reserved.

Intel Software Guard Extensions (SGX) allows applications to run trusted code in a protected environment. For PoET, the trusted code is what ensures that these two requirements are satisfied—keeping the lottery fair.

Proof of Importance (PoI)

The proof of importance (PoI) algorithm is a variation of proof of stake that takes into consideration the role of validators and shareholders in the blockchain operation. These considerations go beyond the size of participant holdings and look at a variety of factors, including, for example, balance, reputation, number of transactions of a single addresses, and more. PoI-based blockchains are expensive to attack, given the computational resources required to do so, and rewards users for contributing to the network security and using the network.

1.5.9 Determining Consensus Needs

It is important to determine consensus needs. When considering the types of blockchain solutions available to an enterprise (some of the more prominent solutions are covered later in this study guide) and determining the true use case for blockchain technology in the enterprise, the following questions may serve as a starting point for determining which solutions to choose:

- How is the process within the enterprise conducted currently, and how will that change with blockchain technology?
- Who are the internal and external parties to the process?
- Do these parties trust each other with data or transactions? Are there levels of trust?
- Who or what validates the accuracy and content of data, and which consensus methodology and blockchain solution meets these validation requirements?
- Are the limited amounts of data to be shared, or will all data be shared?
- Is there a need real time for fast transactions or data processing?

This is a small sample of questions to be asked when seeking the appropriate blockchain solution leveraging the most fitting consensus methodology. Readers are encouraged to have these discussions across the technical and the business units within an enterprise.

1.5.10 Cryptographic Protocols

Cryptographic protocols leveraging proofs are essential to persist any transaction onto a blockchain. Such proofs apply complex algorithms that use the public/private key pair, hashing and hierarchical deterministic keys. The Bitcoin blockchain is used as an example in this explanation cryptographic protocols.

To make generating bitcoins difficult, the hashcash[15] cost-function is employed. Hashcash is the first secure, efficiently verifiable, cost function or proof-of-work function. The benefit of hashcash is that it has no secret keys that have to be managed by a central server or relying party. Thus, hashcash is fully dispersed and infinitely scalable. Hashcash uses symmetric key cryptography, specifically a one-way hashcash function—typically SHA1 or SHA-256. For Bitcoin specifically, hashcash uses SHA256 as the underlying cryptographic hash function.

Hashing takes an input string of any length and provides an output of a fixed length. A cryptographic hash function takes input data of any size and converts it into a hash, in an impossible to reverse or predict way, resulting in a compressed string (for SHA-256, the hash is 32 bytes). Making the slightest change to the input data changes its hash

[15] Hashcash.org, "Hashcash," *www.hashcash.org*

ISACA. All Rights Reserved.

unpredictably, so no party can generate a different block of data that gives exactly the same hash. Thus, by being a fixed compact hash, a party can confirm that it matches only a particular input data set. Thus, blocks on the Bitcoin blockchain can be identified by their hash, which serves the twofold purpose of identification and verification of data veracity.

In addition, this cryptographic hash function requires a difficulty factor, which is accomplished in the code by requiring that this hash output has some quantity of preceding zeros.

A nonce, which means number only used once, is a number that is added to a hashed or encrypted block in the blockchain and is instrumental in setting the mining difficulty level. The nonce is the first number that a blockchain miner must discover before solving for a block in the blockchain and receiving a reward.

Determining which string to use as the nonce requires a significant amount of trial and error, because it is random. A miner on a blockchain network must guess a nonce, append it to the hash of the current block, rehash the value and compare this to the target hash. If the resulting hash value meets the requirements, the miner has found the solution and is awarded the block. Because it is extremely unlikely any miner will successfully guess the nonce on a first attempt, the miner may need to test numerous nonce options before arriving at a correct solution. This is the fundamental reason why so much computer processing power is required to mine bitcoin.

Block difficulty is kept the same across the entire network; all miners have the same chance of arriving at the correct hash. Cryptocurrency networks, at design and coding, typically establish a target number of blocks to be administered in a specified time. The difficulty is adjusted on occasion to ensure that this coded target number is met by the network. If the number of blocks does not meet this target, then the difficulty is reduced to accommodate.

For the most part, blockchain transactions are immutable, meaning that the transactions are unchanging and unable to be changed, over time. Any attempt to change data or a transaction is rejected by the network because it would cause a breakdown in the continuity and progression of data in each block.

Each public and private blockchain has its own set of cryptographic rules and is designed to accomplish a particular task. In addition, there are numerous consensus methodologies and blockchain platforms, some of which are addressed later in this course. The underlying processes of blockchain technology can often be very sophisticated, but the technology represents an innovative way to exchange value between stakeholders.

Figure 1.4 summarizes some core benefits of using blockchain technology.

Figure 1.4—Blockchain Terminology

Term	Definition
Anonymous	Also known as "pseudo-anonymous", in which private information (for a public blockchain) associated with transactions are linked to wallet addresses and public keys, and no personally identifiable information is viewable.
Distributed	More reliable; system is more resilient. Component failure is minimized, and transactions are encrypted and stored on multiple nodes globally.
Decentralized	No need to trust a central authority, which results in less likelihood of a single point of failure, and allowing the blockchain to be censure resistant.
Immutable	Data is append-only and cannot be modified; for the most part, public blockchain transactions are tamperproof.
Transparent	Transparency allows for transaction history to be more easily audited, and offers greater accuracy and consistency of data.

The oldest and currently most prominent use of blockchain technology is cryptocurrency.

ISACA. All Rights Reserved.

1.6 Chapter 1 Knowledge Check

REVIEW QUESTIONS

1. DigiCash was almost bought by _______, but the creator declined the offer.

   A. Apple®
   B. Google®
   C. Amazon®
   D. Microsoft®

2. Satoshi Nakamoto is a well known and established expert in the financial industry.

   A. True
   B. False
   C. Maybe

3. Distributed ledgers are all self-validating.

   A. True
   B. False

4. To encrypt a message with PKI, the sender needs which two keys:

   A. Sender's public key, receiver's public key
   B. Sender's private key, receiver's public key
   C. Sender's public key, receiver's private key
   D. Sender's private key, receiver's private key

5. Bitcoin wallet addresses use:

   A. Letters, numbers and special characters
   B. Base64 encoding
   C. Letters and numbers
   D. Hexadecimal

6. A key rule to blockchain transactions is:

   A. Only primary nodes can initiate or receive a transaction.
   B. Only secondary nodes can initiate or receive a transaction.
   C. Primary nodes can only initiate transactions and secondary nodes can only receive transactions.
   D. Any node on the network can initiate or receive a transaction.

7. Soft forks allow a blockchain system to:

   A. Facilitate backwards compatibility of previous versions on the blockchain
   B. Drastically change the blockchain version and force nodes to upgrade to the latest software
   C. Require that only a few nodes upgrade on the network
   D. Eat spaghetti without scratching their forks on their plate

ISACA. All Rights Reserved.

8. Which of the following is a consensus mechanism requiring each node to wait a random amount of time, and designating the first node to finish that randomized time interval as the leader for the new block.

   A. Practical Byzantine fault tolerance (pBFT)
   B. Proof of elapsed time (PoET)
   C. Proof of importance (PoI)
   D. Proof of work (PoW)

9. ________________ takes into consideration the role of validators and shareholders in blockchain operation.

   A. Practical Byzantine fault tolerance (pBFT)
   B. Proof of elapsed time (PoET)
   C. Proof of stake
   D. Proof of importance

10. A nonce is a:

    A. A nonsense number
    B. Number that is only used once
    C. Blockchain node
    D. Transparent blockchain transaction

Answers on page 26

ISACA. All Rights Reserved.

Chapter 1 ANSWER KEY

Review Questions

1. A. Apple
   B. Google
   C. Amazon
   D. Microsoft. Refer to page 14.

2. A. True
   B. False
   C. Maybe. Refer to page 14.

3. A. True
   B. False. Refer to page 15.

4. A. Sender's public key, receiver's public key
   B. Sender's private key, receiver's public key. Refer to page 16.
   C. Sender's public key, receiver's private key
   D. Sender's private key, receiver's private key

5. A. Letters, numbers and special characters
   B. Base64 encoding
   C. Letters and numbers. Refer to pages 17-18.
   D. Hexadecimal

6. A. Only primary nodes can initiate or receive a transaction.
   B. Only secondary nodes can initiate or receive a transaction.
   C. Primary nodes can only initiate transactions and secondary nodes can only receive transactions.
   D. Any node on the network can initiate or receive a transaction. Refer to page 18.

7. **A. Facilitate backwards compatibility of previous versions on the blockchain. Refer to page 19.**
   B. Drastically change the blockchain version and force nodes to upgrade to the latest software
   C. Require that only a few nodes upgrade on the network
   D. Allow a blockchain system to eat spaghetti without scratching their forks on their plate

8. A. Practical Byzantine fault tolerance (pBFT)
   B. Proof of elapsed time (PoET). Refer to page 21.
   C. Proof of importance (PoI)
   D. Proof of work (PoW)

ISACA. All Rights Reserved.

9.
A. Practical Byzantine fault tolerance (pBFT)
B. Proof of elapsed time (PoET)
C. Proof of stake
D. Proof of Importance. Refer to page 22.

10.
A. A nonsense number
B. Number that is only used once. Refer to page 23.
C. Blockchain node
D. Transparent blockchain transaction

ISACA. All Rights Reserved.

Page intentionally left blank

ISACA. All Rights Reserved.

Chapter 2:

Cryptocurrency

Cryptocurrency

2.1 Learning Objectives 30
2.2 Overview 30
2.3 Types of Virtual Currencies 31
2.4 Chapter 2 Knowledge Check 38

ISACA. All Rights Reserved.

Cryptocurrency

2.1 Learning Objectives

After completing this chapter, learners will be able to:

- Articulate what cryptocurrencies are.
- Distinguish differences among cryptoassets.
- Explain benefits of crypto and digital assets.
- Define smart contracts, their benefits and risk.

2.2 Overview

The term cryptocurrency has, in many minds, achieved global awareness. However, for many, it is still a mystery, conjuring up images of illicit activity, wild speculation ("What price will bitcoin be at the end of the year?" is a common refrain across many ecosystems, particularly financial markets) and confusion. This chapter reviews history and definitions of cryptocurrency, its benefits and its current standing in the world of markets and technology. In addition, virtual currency, cryptocurrency, cryptoasset and other relevant terms are differentiated.

There are dozens of cryptocurrencies, cryptoassets, virtual currencies and more. By May 2020, there were at least 5,500 cryptocurrencies publicly known, spanning more than 22,000 market venues.[16]

Bitcoin is not considered the first cryptocurrency. The concepts of digital currency, cryptocurrency and cryptoassets have been in existence for some time. The idea of electronic cash, or ecash, is almost four decades old.

David Chaum's creation of the anonymous cryptographically based electronic money, i.e., ecash, was followed over a decade later by the creation (and ultimately failure) of DigiCash. Not long afterward, in 1996, the National Security Agency Office of Information Security Research and Technology published "How to Make a Mint: the Cryptography of Anonymous Electronic Cash."[17] This paper portrays a cryptocurrency system, broadly defines the basic concepts of electronic cash and electronic payment systems, and explains the cryptographic electronic cash protocols, considering current authentication mechanisms. The paper also went far deeper into the specific technical and security implementation considerations required for an electronic cash system. With its later republication via MIT and the American Law Review, the paper began to push the idea of e-cash further into the academic and government awareness.

In 1998, Wei Dai, a Chinese computer engineer, published a paper designating b-money, illustrating a distributed, fully anonymous electronic cash system. This system is, in principle, the precursor framework for all current cryptocurrency platforms. He described the system as "a scheme for a group of untraceable digital pseudonyms to pay each other with money and to enforce contracts amongst themselves without outside help."[18]

In the same year, well-known computer scientist (and legal scholar) Nick Szabo designed a means of creating a decentralized digital currency that he named bit gold. Although never put into production, bit gold was the final antecedent of bitcoin, in that any member of the bit gold network was required to dedicate computer power to solving cryptographic puzzles, which increase in complexity over time and require the use of public/private key encryption. In the Szabo model, each solved puzzle was part of the next round of puzzle solving, creating an ever growing, append-only chain of new data and transactions. Like bitcoin and other cryptocurrencies, new blocks were

[16] Coinmarketcap, 20 May 2020, *https://coinmarketcap.com*
[17] Law, L.; S. Sabett; J. Solinas; "How to Make a Mint: The Cryptography of Anonymous Electronic Cash," 18 June 1996, *https://groups.csail.mit.edu/mac/classes/6.805/articles/money/nsamint/nsamint.htm*
[18] Dai, W.; bmoney.txt, *www.weidai.com/bmoney.txt*

ISACA. All Rights Reserved.

verified and time stamped. The majority of the network had to agree to the new block, driven by the solution to the cryptographic puzzle. Less than a decade later, the Satoshi paper for bitcoin was published.

Cryptocurrency history shows an evolution towards greater anonymity and security in value transfer by leveraging the power of peer-to-peer technology and cryptography to create new ways of exchanging value. But, what are the requirements that inform the definition of a cryptocurrency? A multitude of definitions have been put forward, but for the purposes of this study guide, ISACA uses the Lansky definition,[19] which suggests that a cryptocurrency (and its underlying system or blockchain) must satisfy the following six core conditions:

- The system keeps an overview of cryptocurrency units and their ownership.
- The system does not require a central authority, its state is maintained through distributed consensus.
- The system defines whether new cryptocurrency units can be created. If new cryptocurrency units can be created, the system defines the circumstances of their origin and how to determine the ownership of these new units.
- Ownership of cryptocurrency units can be proved exclusively cryptographically.
- The system allows transactions to be performed in which ownership of the cryptographic units is changed. A transaction statement can only be issued by an entity proving the current ownership of these units.
- If two different instructions for changing the ownership of the same cryptographic units are simultaneously entered, the system performs at most one of them.

2.3 Types of Virtual Currencies

Virtual currencies have been in existence for some time, and indeed most people have had some interaction with virtual currencies in their daily lives. In general, a virtual currency (virtual money) is a type of digital representation of value that is not fully regulated in the way that sovereign governments regulate fiat currency. Thus, virtual currencies are digital representations of value, not created or issued by a central bank or sovereign state, which can be used as a method of exchange. Often, these virtual currencies are issued and managed by the developers of their particular ecosystem and are acceptable for use among the participants in that ecosystem.

Given this definition, many examples of such virtual currencies can be found. For example, the multitude of loyalty and reward programs, such as credit card rewards and airline loyalty programs, account for billions of dollars' worth of value in their particular ecosystems. They are commonly known as interoperable within their closed ecosystems, and, in all instances, the value of these reward points is determined wholly and entirely by the organization that created them. For example, members can use airline reward points to book travel with that specific airline or with partner airlines. Conversely, airline partner programs, such as participating credit card programs, can offer members the opportunity to earn airline reward points based on the number and amount of purchases that they make with their credit cards. However, airline rewards points cannot be used as fiat currency. An airline rewards program member cannot appear at a retail store and purchase an item with airline points as a replacement for cash.

Most recently, there has been a rise of virtual currencies for use in popular gaming platforms and ecosystems. For example, V-Bucks, game virtual currency for the popular Fortnite[20] massive multiplayer online game. V-Bucks can be purchased online or earned as part of achieving goals within the game, but again cannot be used for anything other than what the game designers dictate, and not for any goods or services in the real world. There have been growing concerns among international legislators and regulators regarding the use of in-game virtual currencies for a variety of illicit activities, including the stealing of credit card information or money laundering.[21]

[19] Lansky, J.; "Possible State Approaches to Cryptocurrencies," *Journal of Systems Integration*, 2018, *http://si-journal.org/index.php/JSI/article/viewFile/335/325*

[20] Fortnite Wiki, "V-Bucks," *www.fortnite.fandom.com/wiki/V-Bucks*

[21] Gershel, B.; "Popular Video Game Serves as a Reminder of Regulatory Risks of In-game/In-App Virtual Currencies," Money Laundering Watch, Ballard Spahr, LLP, 29 January 2019, *www.moneylaunderingnews.com/2019/01/popular-video-game-serves-as-a-reminder-of-the-regulatory-risks-of-in-game-in-app-virtual-currencies/*

ISACA. All Rights Reserved.

There are two primary types of virtual currency: centralized and decentralized. So far in this section, the examples of the virtual currencies noted that all exist in a centralized fashion. They all have single creators and administrators who not only determine the usage but also any value of these virtual currencies.

A decentralized virtual currency, such as bitcoin, has several distinct characteristics that were noted before but bear repeating. These include:

- Operates on a peer-to-peer basis; virtual currency is not issued by a central party or controlling entity.
- Allows for the secure transferring of unique digital assets from one party to another within the ecosystem, using cryptographic proofs instead of trust.
- Leverages the processing of computer code to solve an increasingly challenging mathematical problem, which generates a unit of virtual currency.
- The system is fully transparent; all completed transactions written to the applicable blockchain are available for viewing.
- Prevents double spending; like physical cash, one note of the cash represents a single unit of that money. Digitally generated tokens of value or virtual currencies were previously prone to duplication, in the same way that one can copy a digital photo and email it to a colleague. The ability of a decentralized virtual currency to make certain that each unit of that virtual currency is truly and perpetually unique solves the problem of double spending.

All participants in the ecosystem, whether they own or transact in units of the virtual currency, are anonymous.

Bitcoin (and all virtual currencies like it) are decentralized, virtual cryptocurrency.

Bitcoin, as designed, has some very specific characteristics, such as:

- There can never be more than 21-million bitcoins.
- Through the process of mining, new bitcoins are produced approximately every 10 minutes on the bitcoin network.
- The mining reward, i.e., the incentive for bitcoin miners to continue to validate transactions on the Bitcoin blockchain, halves at about every 210,000 blocks mined.
- Bitcoin is designed to be a form of electronic money only; it is not meant to serve any other function.

There are other types of this decentralized virtual currency. Since the launch of Bitcoin in 2009, the birth of hundreds of different types of these tokens has occurred. Some of them are described in the following sections.

2.3.1 Altcoins

The definition of altcoin is slightly tricky, because there really is no formal description. *The Wall Street Journal* defined altcoins as "alternative digital currencies,"[22] while some cryptocurrency industry participants consider altcoins as alternatives to bitcoin. This study guide considers altcoin to be all other cryptocurrencies other than bitcoin.

Interestingly, altcoins can diverge from bitcoin and the underlying Bitcoin blockchain in every way, including how the cryptocurrency is distributed, mining methods and consensus methodologies (which are discussed in later sections), to the ability to use the altcoin blockchain for other functions, such as the creation and sharing of decentralized applications.

[22] Yang, S.; "Want to Keep Up with Bitcoin Enthusiasts? Learn the Lingo," *Wall Street Journal*, 31 January 2018, *www.wsj.com/articles/want-to-keep-up-with-bitcoin-enthusiasts-learn-the-lingo-1517394601*

ISACA. All Rights Reserved.

Altcoins arose to address some issues and shortcomings associated with bitcoin originally. Namely, that bitcoin and its blockchain are relatively slow (from a payment processing perspective, bitcoin on average processes transactions into blocks approximately every 10 minutes; Visa can process up to 65,000 transactions per second[23]), fairly computational and energy intensive, and totally public. Many altcoins arose to address one or all of these issues.

Historians of cryptocurrency widely consider Namecoin to be the first altcoin. It launched in 2011 with a goal of decentralizing domain registration (away from the Internet Corporation for Assigned Names and Numbers or ICANN, the centralized nonprofit organization that is in charge of maintaining and coordinating the Internet). The ensuing years since the launch of Bitcoin have seen the creation of thousands of different altcoins, with names such as Ripple,[24] Litecoin,[25] Peercoin,[26] Zcash,[27] all with alternative goals, technologies and values.

2.3.2 Cryptoassets

Generally, those decentralized virtual currencies (and their underlying blockchain technology layers) that are meant to achieve something other than the exchange of value, are grouped in a large segment called cryptoassets. Because blockchain technology in general can accomplish a wide range of functions, it is not surprising that the technology has been used for a variety of broad solutions. For example, some cryptoassets have been created to use cryptography to control the generation of units of an asset and verify their transfer between parties. Others have been developed to act as a commodity or a form of payment, while others are more like securities (i.e., stocks and bonds).

Some broad categories of cryptoassets are defined in the following sections.

Security Tokens

Tokens that are considered a quantity of a specified Investment, including rights to ownership, payment of a specific sum under a contract and entitlement to future profits. For the most part, the majority of countries, transnational organizations and regulatory agencies consider these some form of securities that are subject to the relevant securities laws and regulations within those jurisdictions. The issuers of Initial Coin Offerings (ICOs)[28] throughout 2017 and 2018 raised billions of dollars in value around the world to launch new technology platforms, sometimes on nothing more than a basic presentation document. These issuers quickly ran afoul of these regulators, when it was adjudicated that many of these issuances amounted to unregistered securities sales, and were illegal. (Legal action is still being undertaken around the world against ICO issuers, at the time of this writing.)

Utility Tokens

Utility tokens are digital assets used to finance the creation of a network by providing its buyers with a pledge of being able to use some network ecosystem or products. They do not provide any token holder a legal or economic right of ownership over the developer nor any part of the ecosystem. Although usually based on an underlying blockchain platform, utility tokens are not mined for their creation, but are created by the developers. Any value they may have is based solely on their inherent functions within the relevant ecosystem. For example, utility token holders may be given a specific token to use a company's products or services. A good example is the basic attention token (BAT),[29] which was created to give token holders the ability to participate in a decentralized advertising marketplace connecting publishers, advertiser and users. Utility tokens cannot lawfully be traded on a secondary

[23] VISA, "VISA Fact Sheet: What you need to know about one of the world's largest payment companies," *https://usa.visa.com/dam/VCOM/download/corporate/media/visanet-technology/aboutvisafactsheet.pdf*
[24] Ripple, *https://ripple.com*
[25] Litecoin, *https://litecoin.org*
[26] Peercoin, *https://www.peercoin.net*
[27] Zcash, *https://z.cash*
[28] Frankenfield, J.; "Initial coin offering," *Investopedia*, 3 November 2020, *www.investopedia.com/terms/i/initial-coin-offering-ico.asp*
[29] Basic Attention Token, *https://basicattentiontoken.org*

ISACA. All Rights Reserved.

market or for cash (except for launch), because this would most probably run afoul of securities or money servicing laws around the world.

Stablecoins

A relatively new entrant to the world of blockchain and cryptoassets, stablecoins are cryptoassets created to address the problem of volatility in the cryptoasset world. Due to relatively low liquidity compared to other instruments and unresolved global regulatory concerns, bitcoin (for example) has proven to show double digit percentage volatility, sometimes on a daily basis. By tying, or pegging, the cryptoasset to some underlying asset or group of assets, minimization of price volatility can be achieved and potentially wider adoption. There are several types of stablecoins, including but not limited to:

- Stablecoins pegged to fiat currency, such as Trust Token (TrueUSD)[30] or Circle (USDC)[31]
- Adjustable peg stablecoins, which are allowed to fluctuate between the par value and an acceptable range of predetermined rates
- Commodity-backed stablecoins pegged to the price of a barrel of oil, an ounce of gold or some other commodity
- Stablecoins pegged to a basket of bank deposits and short-term government securities. Facebook Libra may be a good example of this.

There are a multitude of complexities involved in this ever-expanding segment of cryptoassets, including the issues of pegging definition, collateralization and redeemability.

2.3.3 Cryptocurrency Benefits

The broad benefits of cryptocurrencies and cryptoassets include the following:

- Ability to achieve faster and cheaper payments within country and importantly internationally, because cross-border payments and remittances tend to be very expensive, slow and error prone
- Removing the need for traditional intermediaries, due to the decentralized nature of cryptocurrencies, such as bitcoin, can significantly reduce or eliminate the cost and friction of transactions
- The potential to create new types of payments and value exchange models, such as the ability to offer small or even micropayments, which can allow for significant aid to the global unbanked and underbanked[32]
- Greater speed and availability of transaction processing, including instantaneous clearing and settlement of securities transactions, which tend to be relatively costly and friction filled
- The tokenization of real and digitally native securities, such as stock or real estate, offer increased liquidity, allow for cheaper and faster trade settlement and provide for much greater transparency regarding the history of an asset from the moment of its creation

2.3.4 Smart Contracts

Smart contracts are one of the more prominent and, in some instances, controversial capabilities to evolve from the development and implementation of blockchain technology over the past several years. The definition of contract according to Merriam Webster, which is "a binding agreement between two or more persons or parties; especially… one that is legally enforceable."[33] Conventionally, contracts are produced in writing, leaving the function and role of contract performance and enforcement to the actual parties of the contract, or if necessary, outside legal enforcement

[30] TrustToken, "TrueUSD," *www.trusttoken.com/trueusd*
[31] Circle, "USDC: the fastest growing, fully reserved digital dollar stablecoin," *www.circle.com/en/usdc*
[32] Levine, A.B.; "Banking the Unbanked: How the Crypto Community Can Make a Big Impact," Coindesk, 30 April 2020, *www.coindesk.com/banking-the-unbanked-how-the-crypto-community-can-make-a-big-impact*
[33] Merriam-Webster, "contract," *www.merriam-webster.com/dictionary/contract*

ISACA. All Rights Reserved.

entities, such as arbitrators or courts. Human interaction and errors can leave fairly wide room for problems, including misinterpretation or potentially incorrect or absent performance. The result can be very costly to all parties and can lead to enforcement disputes.

The proposed definition of smart contract by Nick Szabo, set forth several years ago is: "a set of promises, specified in digital form, including protocols within which the parties perform on these promises."[34]

Much of the discussion regarding smart contracts in the ensuing years is predicated on this early definition. Smart contracts seek to take the covenants and verbiage of traditional physical contracts and convert them into a digital form so that they maintain their compliance requirements and can be fully executed and enforceable.

From a technology perspective, a smart contract consists of a computer program and a transactional protocol, which are designed to automatically execute, control and/or document respectively legally relevant events and actions according to the terms of a contract, of an agreement or of a negotiation.[35]

But for what reason are smart contracts needed? What are they meant to accomplish? When the challenges and costs associated with legal contract creation and enforcement are considered, using technology to evolve how contracts are created and used can be envisioned. Thus, smart contracts are intended to take advantage of the potential of blockchain technology and automatic processes to provide authenticated contract creation and execution on a peer-to-peer network, without the need for independent verification or third-party management. Smart contracts that live on a blockchain are meant to be self-enforcing. This is clearly far beyond the legacy of contracts written on paper, with lawyers, in traditional language. In coding a smart contract, the covenants of the agreement can be automated for a binding agreement and, on the meeting of certain conditions, with all transactions occurring because of that contract being written to a blockchain in an immutable, secure way.

More technically, computer code can serve the function of defining terms of a contract and those events that initiate certain activities. For example, Joseph Dewey proposed the following simple example of a promise to pay in 2016,[36] coded in JavaScript:

Example:

```
var promisePay= {

    text: 'hereby promises to pay to the order of',

    call: function (maker, payee) {

    var reply = maker + ' ' + this.text + ' ' + payee;

    return reply;

    }

};
```

[34] Szabo, N.; "Smart Contracts: Building Blocks for Digital Markets,"1996, *http://www.fon.hum.uva.nl/rob/Courses/InformationInSpeech/CDROM/Literature/LOTwinterschool2006/szabo.best.vwh.net/smart_contracts_2.html*
[35] Mearian, L.; "What's a smart contract (and how does it work)?," Computerworld, 29 July 2019, *www.computerworld.com/article/3412140/whats-a-smart-contract-and-how-does-it-work.html*
[36] Dewey, J.; "What if we developed legal contracts like we developed software applications?," Medium, 6 April 2016, *https://medium.com/@ConsenSys/what-if-we-developed-legal-contracts-like-we-developed-software-applications-1b5bc8fbb915*

ISACA. All Rights Reserved.

This simple coding defines a property and a two-argument function, namely the maker and the payee. As this function is called and the proper maker/payee information is entered, "the object will return the text "{Maker} promises to pay to the order of {Payee}".[37]

Therefore, the following line of code:

'promisePay.call('Bob', 'John');',

will return the following to the user:

Bob hereby promises to pay to the order of John.

Figure 2.1 graphically represents the process of smart contracts.

Figure 2.1—Applying Business Logic With Smart Contracts

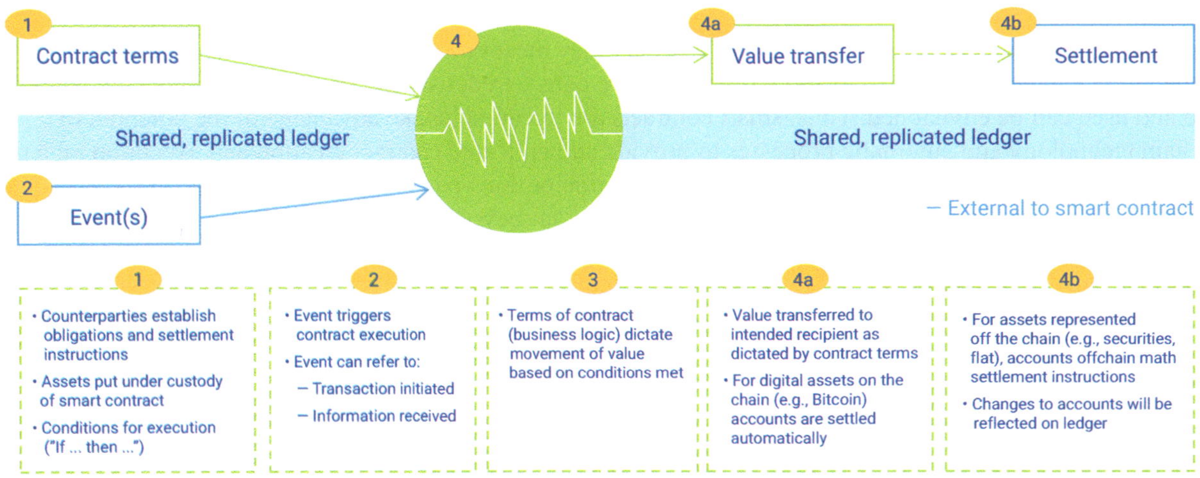

Source: BBVA Research, "Smart contracts: the ultimate automation of trust?," Digital Economy Outlook, October 2015, *www.bbva.com/wp-content/uploads/en/2016/11/Digital_Economy_Outlook_Oct15_Cap1.pdf*

Most recently, smart contracts have found singular interest in financial institutions, given the volume and frequency of contracts that financial firms enter into and execute on behalf of themselves and/or their clients. Applying the blockchain to smart contracts so that data can automatically be verified, and terms executed, offers the potential for much more accurate and less costly contract management, all secured with the benefits of blockchain technology. Some specific implementations of smart contracts in financial markets include:

- Securing and processing of loans on a blockchain, along with collateral management and ownership. In this example, a missed loan payment can result in the automatic revocation of the borrower's digital keys, essentially locking them out of the collateral.
- The creation and management of escrow accounts that automatically and securely oversee any exchange between parties. A buyer of goods, for example, can transfer remittance to the smart contract. After an internal trigger is activated (for example GPS services that notify of goods delivery), the smart contract automatically processes payment.
- Securities, such as equities or bonds whose purchase or sale is defined by smart contracts, and transactions that can not only be automated, but also allow for instantaneous transfer of ownership and payment. Smart contracts on blockchain allow for the use of cryptographic keys to represent the asset in digital form. They can also

[37] *Ibid.*

ISACA. All Rights Reserved.

monitor the performance of digital or nondigital assets and also be used for a variety of derivative instruments, such as options or futures.

Although financial markets is the use case here, using smart contracts is possible over a wide swath of industries, including health care, digital media, supply chain and more.

Despite all the benefits of smart contracts, there is risk associated with using computer code as a means to supplement or replace the usual methods of contract creation. This risk and commensurate challenges are part of the headwinds to wider adoption of smart contracts.

Real-world implementation of contracts requires a degree of flexibility, and legacy contracts are often imprecise by design. Given the immutability of blockchain, smart contracts need to have coding and mechanisms that allow for the amendment of contracts terms if and when needed.

Given that smart contracts are software, they can administer and enforce the terms of contracts and state of data between parties. Although many jurisdictions allow for the legality of smart contracts,[38] the ability to rely solely on smart contracts for contract creation and enforcement may differ across markets and jurisdictions. For example, in the United States, many but not all states adhere to the Uniform Electronic Transactions Act (UETA), which some consider to adequately cover smart contacts. Care should be taken to consult legal counsel when trying to understand the enforceability of smart contracts.

Like with blockchain adoption, the true benefits of smart contracts will be recognized with wider scale adoption and usage. When a wide number of commercial entities begin to leverage blockchain and smart contracts, the network effects of this will allow for more robust usage, revolution and development. However, as of the time of this writing, smart contracts are still struggling to develop wider usage.

The landscape of regulation associated with smart contracts is unclear. Smart contracts and blockchain are proving a challenge to many regulators around the world, because, in particular, public blockchains make enforcement and targeting of specific commercial entities difficult. Thus, the liability associated with these challenges may dissuade some entities from using smart contracts until legislation and regulation in their respective jurisdictions are more clearly defined.

The Decentralized Autonomous Organization (DAO) is one of the more prominent demonstrations of smart contract and blockchain capability. The DAO is a computer program that existed on the Ethereum Blockchain using smart contracts. In 2016, the DAO was launched as a form of venture capital funds and crowdfunding campaign, with a mission to provide a new decentralized business model for establishing commercial and nonprofit businesses.[39] It had no board nor management team, was stateless (i.e., not associated with any particular jurisdiction) and the code was fully open source. It was one of the largest crowdfunding campaigns in history, raising over 168-million dollars, in exchange for tokens that were freely transferable and allocated voting rights and ownership. The DAO existed to invest in ideas, companies and innovative projects, and sought to provide positive returns to token holders.

On June 17, 2016, a weakness in the smart contract code of the DAO was compromised when a flaw in the code was exploited by malicious actors. Not long afterwards, the US Securities and Exchange Commission (SEC) released a report in which it found that the tokens issued by the DAO constituted securities under the federal securities laws. To solve for this issue, the Ethereum community invoked a hard fork of Ethereum, creating a second Ethereum blockchain and cryptocurrency, and moving the funds raised to a new wallet where they can be exchanged back by their original owners. This was a complicated tale at the regulatory, business and technology layers. The incident had a cooling effect on smart contract usage and clearly highlighted some challenges of smart contracts.

[38] Finextra, "UK charges ahead with legal certainty for smart contracts and cryptoassets," 18 November 2019, *www.finextra.com/newsarticle/34790/uk-charges-ahead-with-legal-certainty-for-smart-contracts-and-cryptoassets*

[39] Wikipedia, "The DAO (organization)," *https://en.wikipedia.org/wiki/The_DAO_(organization)*

ISACA. All Rights Reserved.

2.4 Chapter 2 Knowledge Check

REVIEW QUESTIONS

1. Virtual currencies are:

   A. Virtualized versions of fiat currency
   B. Regulated by a centralized bank
   C. Managed by governments
   D. A digital representation of value

2. An important aspect of a decentralized virtual currency is:

   A. Having a single point of failure for attribution
   B. Ensuring that unit worth is variable
   C. Transparency
   D. Disrupting current markets

3. One of the first AltCoins was:

   A. Bitcoin
   B. Dogecoin
   C. Namecoin
   D. Litecoin

4. A type of cryptoasset that represents a quantity of a specified investment, including rights to ownership, payment of a specific sum under a contract or entitlement to future profits is known as a:

   A. Stablecoin
   B. Cryptocurrency
   C. Utility token
   D. Security token

5. This cryptoasset type does not provide any token holder a legal or economic right of ownership:

   A. Stablecoin
   B. Cryptocurrency
   C. Utility token
   D. Security token

6. One benefit of using cryptocurrencies is that:

   A. They are considered more stable than most FIAT currencies.
   B. Payments are faster and cheaper than conventional methods.
   C. They add intermediaries for oversight.
   D. They are unable to split transactions into micropayments.

7. Smart contracts are legally binding documents.

   A. True
   B. False

ISACA. All Rights Reserved.

8. Smart contracts have many uses and advantages, but one of the most advantageous reasons to utilize smart contracts over traditional contracts is that:

   A. They are easy to write and anyone can do it.
   B. Smart contracts feature automation of processes.
   C. Once in the blockchain, they can be easily modified or changed.
   D. Their regulation is straightforward and well documented.

Answers on page 40

ISACA. All Rights Reserved.

Chapter 2 ANSWER KEY

Review Questions

1. A. Virtualized versions of fiat currency
   B. Regulated by a centralized bank
   C. Managed by governments
   D. A digital representation of value. Refer to page 31.

2. A. Having a single point of failure for attribution
   B. Ensuring that unit worth is variable
   C. Transparency. Refer to page 32.
   D. Disrupting current markets

3. A. Bitcoin
   B. Dogecoin
   C. Namecoin. Refer to page 33.
   D. Litecoin

4. A. Stablecoin
   B. Cryptocurrency
   C. Utility token
   D. Security token. Refer to page 33.

5. A. Stablecoin
   B. Cryptocurrency
   C. Utility token. Refer to page 33.
   D. Security token

6. A. They are considered more stable than most FIAT currencies.
   B. Payments are faster and cheaper than conventional methods. Refer to page 34.
   C. They add intermediaries for oversight.
   D. They are unable to split transactions into micropayments.

7. A. True
   B. False. Refer to pages 35-37.

8. A. They are easy to write and anyone can do it.
   B. Smart contracts feature automation of processes. Refer to page 36.
   C. Once in the blockchain, they can be easily modified or changed.
   D. Their regulation is straightforward and well documented.

ISACA. All Rights Reserved.

Chapter 3:

Topology and Blockchain Models

Topology and Blockchain Models

3.1 Learning Objectives 42
3.2 Overview 42
3.3 Public Blockchains 42
3.4 Private Blockchains 43
3.5 Hybrid Blockchains 44
3.6 Model Comparisons 45
3.7 Chapter 3 Knowledge Check 46

ISACA. All Rights Reserved.

Topology and Blockchain Models

3.1 Learning Objectives

After completing this chapter, learners will be able to:

- Identify different types of blockchain models and topologies.
- Describe benefits and challenges of public blockchain implementation.
- Explain uses of private blockchain.
- Detail pros and cons of using a hybrid blockchain model.
- Distinguish and compare blockchain models.

3.2 Overview

This chapter examines deeper the topology of blockchain, the types of blockchain models that exist and some potential future models that may emerge based upon growing usage of blockchain across multiple industries and functions.

Topology—or more appropriately computer network topology—is "the arrangement with which computer systems or network devices are connected to each other. Topologies may define both physical and logical aspect of the network. Both logical and physical topologies could be same or different in a same network."[40]

Historically, there have been a variety of computer network topology models, such as point-to-point, bus, star, hierarchical tree, mesh and more.[41] Many are familiar models. Blockchain is based on peer-to-peer (P2P) topology, with all transactions arranged by time and current state of system. Peer-to-peer networking is "distributed application architecture that partitions tasks or workloads between peers."[42] Peers are (for the most part) equally privileged participants in the network, forming a peer-to-peer network of nodes. Peers make a portion of their resources, such as processing power, directly available to other network participants, with no central coordination necessary, and the P2P model allows for no single point of failure, because other nodes can perform the functions needed by the network, if one node fail.

In this context then, the following blockchain (or blockchain topology) types can be considered, starting with the public, open blockchain.

3.3 Public Blockchains

The foremost examples of public blockchains to date are the Bitcoin blockchain and the Ethereum blockchain (which will be discussed in more detail later), though there are quite a few others.

Public blockchains are permissionless by design. In other words, any person or party can join this type of blockchain network; participate as a consensus participant or miner; produce applications, smart contracts or transactions; and have a transparent view to the history of all transactions that have occurred on that blockchain. Participation simply requires the download of the particular blockchain platform open-source software. For computationally intensive blockchains like bitcoin, participants need the appropriate level of available memory and processing power to have a reasonable chance of success at mining and receiving the incentive reward of the network.

[40] Tutorialspoint, "DCN – Computer Network Topologies," *www.tutorialspoint.com/data_communication_computer_network/computer_network_topologies.htm*
[41] *Ibid.*
[42] Wikipedia, "Peer-to-peer," *https://en.wikipedia.org/wiki/Peer-to-peer*

ISACA. All Rights Reserved.

As a decentralized network, no single entity can control the network, but there are relevant concerns about concentration of network processing power, such as in a 51% Attack,[43] known to be a concern for all blockchains that are based on a proof-of-work consensus methodology. This type of attack is by a group of blockchain miners who control more than 50 percent of the network mining hash rate or computing power. Once attained, these attackers can conceivably prevent new transactions from being confirmed, thus allowing them to halt transactions between network participants of even reverse transactions, allowing for the double spending of bitcoin. Although this limit has been approached several times in bitcoin history, the incentivization model is such that bitcoin price often drops as other miners lose incentive to participate, hence, making it unprofitable for the miners.

Public blockchains operate and are secured by the combination of cryptographic verification (e.g., proof of work, the consensus methodology used by bitcoin) and cryptoeconomics—the combination of economic-based incentives to maintain and manage the blockchain ecosystem. (The consensus methodologies are reviewed in more detail later in this study guide.)

Prominent examples of public blockchains include:

- **Bitcoin**—A decentralized digital currency that exists without a central banker or separate administrator. Bitcoins can be sent from user to user on the peer-to-peer bitcoin network without any intervention by an intermediary of third party, with transactions being verified by network nodes through cryptography and logged in a public distributed ledger. Bitcoin uses proof-of-work consensus methodology.
- **Ethereum**—An open-source, public, blockchain-based distributed computing platform and operating system featuring smart contract functionality. It supports a modified version of the original Bitcoin blockchain consensus methodology via transaction-based state transitions. Ethereum also uses proof-of-work consensus; however, as of the date of this writing, it is working to switch to proof of stake.[44]
- **Litecoin**—A peer-to-peer cryptocurrency and open-source software project. Like Bitcoin, the creation and exchanges of tokens is based on an open-source cryptographic protocol, with no central authority management. It was one of the first altcoins, launching in 2011. Although almost identical to Bitcoin, it is designed to process blocks to the blockchain much more quickly (approximately every 2.5 minutes, versus approximately every 10 minutes for Bitcoin). Litecoin uses a slightly modified version of proof of work, using a more complicated algorithm than Bitcoin.[45]
- **Stellar**—An open-source, decentralized payment protocol that allows for fast, cross-border transactions between any pair of currencies. Transactions that take place on the Stellar network are added to the Stellar Blockchain. Stellar uses a consensus methodology known as federated Byzantine agreement (FBA).[46]

3.4 Private Blockchains

As blockchain knowledge grew and proof—of-concept implementations began to emerge across the global enterprise landscape in industries as diverse as financial services, healthcare, supply chain and more, some limiting factors of fully open, public blockchains like Bitcoin began to become apparent. In particular were the rising concern about anonymity (or pseudo anonymity) of user data, and the maintenance of privacy of user information. This is notably true when considering the litany of regulations in financial markets (the Bank Secrecy Act [BSA][47] and the Foreign Account Tax Compliance Act [FATCA],[48] to name just two) or in healthcare (Health Insurance Portability and

[43] Puckrin, N.; "51% Attacks: The Deathly Scenario for any Blockchain," *Infosecurity Magazine*, 12 September 2018, *www.infosecurity-magazine.com/opinions/51-attacks-blockchain/*
[44] Ethereum, *https://ethereum.org/en/eth2/*
[45] *Op cit* Litecoin
[46] Mazieres, D.; "The Stellar Consensus Protocol: A Federated Model for Internet-level Consensus," *www.stellar.org/papers/stellar-consensus-protocol*
[47] Bank Secrecy Act (BSA), Office of the Comptroller of Currency, *www.occ.treas.gov/topics/supervision-and-examination/bsa/index-bsa.html*
[48] IRS, "Foreign Account Tax Compliance Act (FATCA)," *www.irs.gov/businesses/corporations/foreign-account-tax-compliance-act-fatca*

ISACA. All Rights Reserved.

Accountability Act [HIPAA]).[49] Such regulations provide steep penalties to enterprises failing to adhere to the data privacy and other requirements that they mandate. It is against this backdrop that the private blockchain emerged.

Private blockchains are permissioned. Thus, they are designed and operate based on access controls that can restrict which, and what type of, parties can participate, and their functions and capabilities in the blockchain workflow. This is in contrast to public blockchain, in which participants are untrusted and are pseudonymous. Participants in private blockchain are typically known to the operators and inherently considered as trusted participants.

In private blockchains, one or more stakeholders or administrators control the network, introducing a requirement for third-party acceptance to transact on that blockchain, which does not exist in public blockchain implementations. In this context of private blockchain, only those parties participating in a transaction have knowledge about it. For the most part, other parties cannot have access, although many private blockchains allow for layers of different types of permissions (submit data, confirm data, view-only, etc.). Thus, although private blockchains can and usually do have a consensus methodology associated with them, trust still plays an important role in private implementations. Also, a private blockchain does not necessary use or have a native token as an incentive to the miners to support security of the network, which is typical in a public blockchain.

Fairly widely known examples of a private blockchain are implementations of Hyperledger Fabric.[50] Produced as modular, open-source software by the Linux Foundation, it allows users to download the software and cater its functions and system permissions to specific use cases, such as financial record keeping, shipping and supply chain logistics. Hyperledger is an umbrella project of open-source blockchains and related tools, started in 2015 by Linux Foundation, with prominent technology company support, notably IBM and Intel.

A subset of private blockchains is the consortium blockchain, which can be considered a unique blend of public and private blockchain. Most of the difference resides in the consensus methodology. For example, a consortium blockchain may designate a group of mutually agreed validators to administer the network and validate transactions. This clearly differs from public blockchain (anyone can participate) or private blockchain (with one central authority or administrator), and levels of transparency and access are managed conjointly. For the most part, consortia blockchain make the most sense in those ecosystems where multiple stakeholders or firms participate in the same industry. There are some prominent examples in the food shipment or construction industries.

3.5 Hybrid Blockchains

The increasing interest in defining potential use cases for blockchain technology has led to the evolution of what are now known as hybrid blockchains. These are blockchain implementations in which one component is a fully public blockchain (with all the benefits and challenges) and the parallel component is a permissioned private blockchain, allowing for enterprise-level transactions that maintain regulatory compliance and access permissioning. Hybrid blockchains seek to realize the best benefits of public and private blockchain technology, including not being fully open to any participant, while maintaining data immutability and transparency. In a hybrid blockchain, although transactions are not public, they are always available to be verified when necessary. There are some benefits to hybrid blockchains, including:

- Allows for control of the blockchain ecosystem; chosen administrators define who may join and their level of access.
- Privacy protection, while allowing for real-world interaction and access to data.
- Generally lower transaction costs, because fewer validators are needed to process transactions and add blocks.

[49] Centers for Medicare & Medicaid Services, "HIPPA and Administrative Simplification," *www.cms.gov/Regulations-and-Guidance/Administrative-Simplification/HIPAA-ACA/index*

[50] Hyperledger Fabric, *https://hyperledger-fabric.readthedocs.io/en/release-2.0/blockchain.html*

ISACA. All Rights Reserved.

A good current example of hybrid blockchain implementation is the IBM Food Trust,[51] which is designed to improve the efficiency of the global food supply chain, and includes a variety of stakeholders, from distributors, to logistics firms, to farmers and beyond. Walmart is a prominent participant in the IBM Food Trust and mandated, via its Walmart Food Traceability Initiative,[52] that a growing number of its suppliers be a part of this blockchain solution.

3.6 Model Comparisons

Following is a summary of the similarities and differences between public and private blockchain:

- Similarities of public and private blockchains
    - Each node possesses a complete copy of the blockchain data or ledger.
    - Distributed and decentralized over a peer-to-peer network.
    - All data is immutable; both types of blockchain act on an append-only ledger; data records can be added but cannot be deleted or changed.
    - Each rely on numerous nodes or users to validate transactions and additions to the ledger.
- Differences between public and private Blockchains
    - For a public blockchain, anyone can take part in verifying data and adding blocks; only authorized participants can do so in a private blockchain.
    - Public blockchains are decentralized; private blockchains for the most part have greater centralization of management and control.
    - Public blockchain can process fewer transactions in a given time frame than private blockchains, given the consensus methodology work (e.g., proof of work) spread across a potentially global peer-to-peer network.
    - Although there are other considerations, such as network design and capital cost, all things being equal, private blockchains can scale more quickly and often need fewer nodes to manage data validation and block creation.
    - Public blockchains are trustless; private blockchains require that there be permissioned access to share data with trusted parties within the network.
    - Decentralization and size of network allow public blockchains to be more secure and less prone to hacks by bad actors.
    - Public blockchain requires far higher energy consumption, given the significant resources consumed in public blockchain consensus and global peer-to-peer networks.

From a technology implementation perspective, when considering public versus private blockchain solutions, it is important to fully recognize the possible business functions and requirements needed, considering available blockchain capabilities. In addition, enterprise implementation of blockchain requires a thorough understanding of existing technology infrastructure and how the enterprise may interact or even be displaced by blockchain technology. Some useful questions to consider include but are not limited to:

- What problem is being solved? What is the true use case, and does blockchain technology offer a potential viable platform to engage that use case?
- Considering the use case, does a public or private blockchain solution make more sense?
- What are the potential legal and regulatory requirements for any solution for the use case (such as privacy, auditability, regulatory reporting) and how might these be impacted by a blockchain solution?
- Who will define permissioning and governance for the blockchain solution, and how programmatically will that be enforced and managed?

[51] IBM, "IBM Food Trust," *www.ibm.com/blockchain/solutions/food-trust*

[52] Walmart, "Blockchain Supplier Letter," 24 September 2018, *https://corporate.walmart.com/media-library/document/blockchain-supplier-letter-september-2018/_proxyDocument?id=00000166-088d-dc77-a7ff-4dff689f0001*

ISACA. All Rights Reserved.

- Who will participate, and how will these participants be chosen?
- How much effort and resource (e.g., time and human capital) will be required to implement any blockchain solution?
- Is there enough understanding of blockchain within the organization to successfully develop and implement a blockchain solution?
- Will more evolved applications of blockchain be needed, such as those blockchains that allow for the development of an application layer, or those that use smart contracts?

Formal implementation of a blockchain solution requires program office and project management engagement, after management buy-in is achieved. In addition, some of the items on this list can be moved into a discussion about smart contracts.

3.7 Chapter 3 Knowledge Check

REVIEW QUESTIONS

1. One benefit of a decentralized network is that no single entity can control the network.

   A. True
   B. False

2. Public blockchains are __________ by design.

   A. Cryptocurrency oriented
   B. Permissionless
   C. Owned by a single entity
   D. Permissioned

3. This public blockchain has smart contract functionality built in.

   A. Bitcoin
   B. Namecoin
   C. Ethereum
   D. Dogecoin

4. As applied to blockchains, permissioned means that:

   A. CTO approval is required before implementation.
   B. Privatization of the blockchain must be completed.
   C. Blockchains are designed to function based on access controls.
   D. Blockchains distribute tokens for completing computations.

5. A widely known example of private blockchain is:

   A. IBM Hyperledger
   B. Ripple XRP
   C. The Ethereum network
   D. Linux

ISACA. All Rights Reserved.

6. IBM Food Trust™ implements this blockchain solution.

   A. Private
   B. Public
   C. Hybrid
   D. Smart contracts

7. One similarity between public and private blockchain models is that:

   A. They are distributed and decentralized over a peer-to-peer network.
   B. One node contains a full copy of the ledger while other nodes only contain partial ledger copies.
   C. Both are centralized and controlled by a single entity.
   D. There are no similarities.

8. Any enterprise blockchain implementation requires:

   A. Mining nodes
   B. Centralization
   C. Management buy-in
   D. Smart contracts

Answers on page 48

ISACA. All Rights Reserved.

Chapter 3 ANSWER KEY

Review Questions

1. **A. True. Refer to 42.**
   B. False

2. A. Cryptocurrency oriented
   B. Permissionless. Refer to page 42.
   C. Owned by a single entity
   D. Permissioned

3. A. Bitcoin
   B. Namecoin
   C. Ethereum. Refer to page 43.
   D. Dogecoin

4. A. CTO approval is required before implementation.
   B. Privatization of the blockchain must be completed.
   C. Blockchains are designed to function based on access controls. Refer to page 44.
   D. Blockchains distribute tokens for completing computations.

5. **A. IBM Hyperledger. Refer to page 44.**
   B. Ripple XRP
   C. The Ethereum Network
   D. Linux

6. A. Private
   B. Public
   C. Hybrid. Refer to page 45.
   D. Smart contracts

7. **A. They are distributed and decentralized over a peer-to-peer network. Refer to page 45.**
   B. One node contains a full copy of the ledger while other nodes only contain partial ledger copies.
   C. Both are centralized and controlled by a single entity.
   D. There are no similarities.

8. A. Mining nodes
   B. Centralization
   C. Management buy-in. Refer to page 46.
   D. Smart contracts

ISACA. All Rights Reserved.

BLOCKCHAIN
FUNDAMENTALS
An ISACA Certificate

Chapter 4:

Blockchain IT Infrastructure

Blockchain IT Infrastructure

4.1 Learning Objectives 50
4.2 Overview 50
4.3 Blockchain Infrastructure 50
4.4 Chapter 4 Knowledge Check 61

ISACA. All Rights Reserved.

Blockchain IT Infrastructure

4.1 Learning Objectives

After completing this chapter, learners will be able to:

- Describe network topologies that enable blockchain.
- Distinguish different types of nodes on a blockchain network.
- Explain the need for interoperability of blockchain networks.
- Identify the need for application programming interfaces (APIs) within a blockchain system.
- Describe specialized hardware that can be utilized in a blockchain.

4.2 Overview

This chapter provides a deeper review of blockchain IT infrastructure. It reviews the fundamentals of this infrastructure at a holistic level, while referencing noteworthy examples of different blockchain implementations. This chapter also covers nodes, network topology, interoperability and a review of the criteria and challenges for public and private implementations.

4.3 Blockchain Infrastructure

Figure 4.1 shows the difference between centralized, distributed and decentralized infrastructures.

Figure 4.1—Blockchain Infrastructures

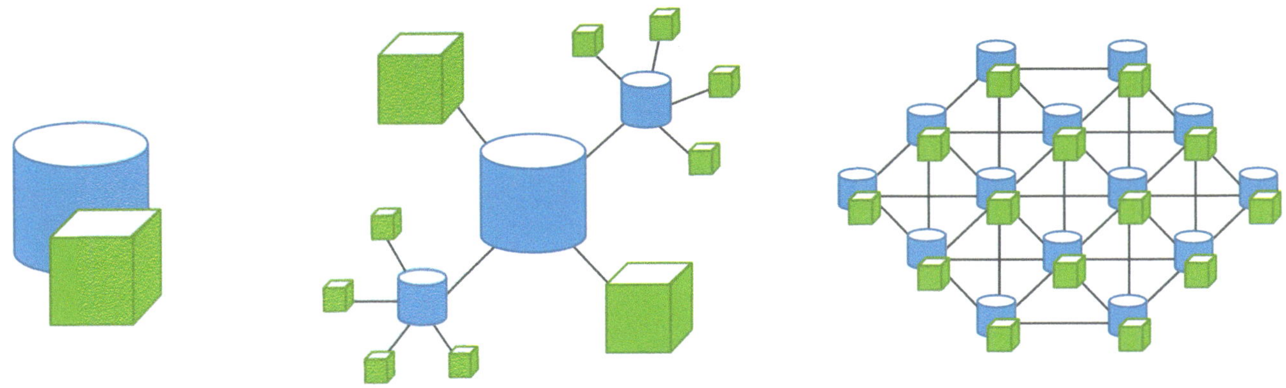

Centralized
(One node does everything)

Distributed
(Nodes distribute work to sub-nodes)

Decentralized
(Nodes are only connected to peers)

Source: StackExchange, "Questions on the terms 'distributed' and 'decentralized,'" *https://ethereum.stackexchange.com/questions/7812/question-on-the-terms-distributed-and-decentralised*. This figure is available under the *Creative Commons Attribution 4.0 International Public* license.

Figure 4.1 points out that distributed systems are a model in which components located on networked computers coordinate actions and process via message passage. In this context, a decentralized system is also a distributed

ISACA. All Rights Reserved.

system.[53] The key point to keep in mind is that decentralization requires no central point of control. Large-scale and widely used Internet applications are distributed, but most are centralized because a single company or entity controls them.[54] Therefore, blockchain-based public platforms like Bitcoin or Ethereum are distributed, decentralized systems that differ only in their use case and consensus methodology.

From a logical process perspective, blockchain functions as a chain of data blocks containing specific data sets in a cryptographically secured way, within a network of participants (nodes), across a decentralized network. Blockchain, therefore, is not based on a client-server architecture, but peer-to-peer network architecture, represented in **figure 4.2**.

Figure 4.2—Client Server and Peer-to-Peer Network Architecture

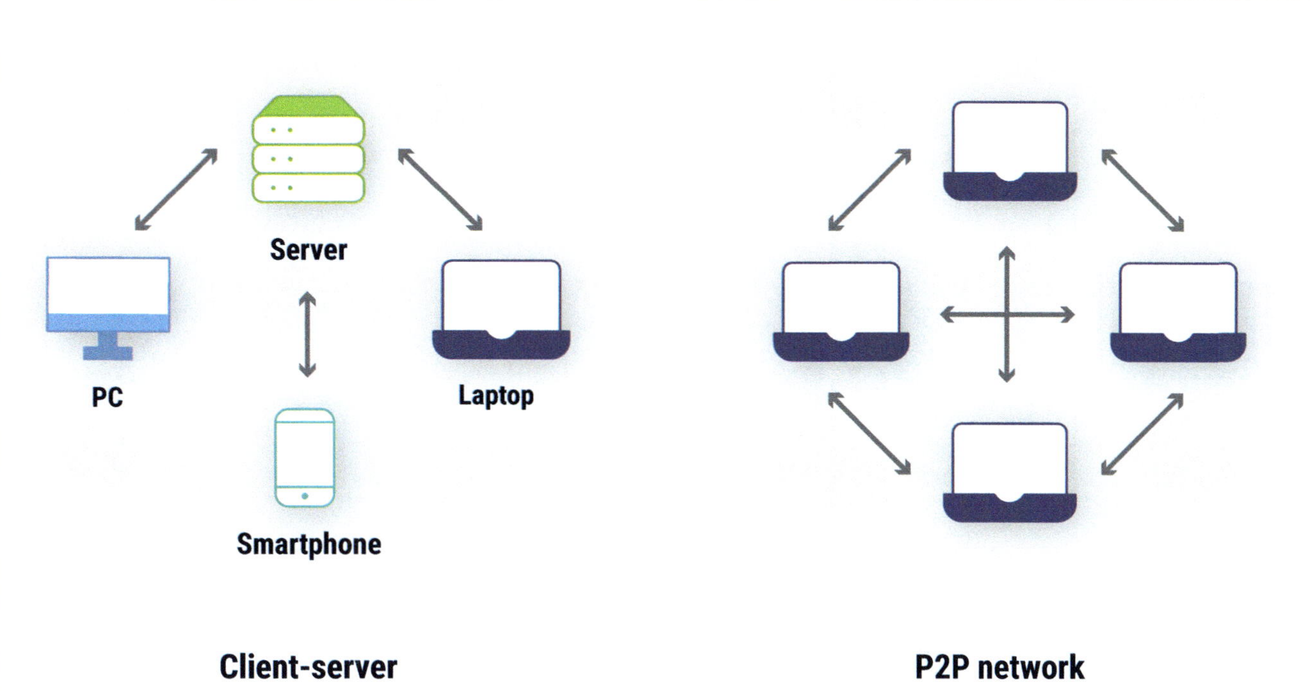

Source: Lastovetska, A.; "Blockchain Architecture Basics: Components, Structure, Benefits & Creation," MLSDev, 31 January 2019, *https://mlsdev.com/blog/156-how-to-build-your-own-blockchain-architecture*

The current architecture of the Internet uses a client-server networking, with centralized databases controlled numerous permissioned administrators. The distributed, decentralized nature of blockchain allows for every participant in the network (with some exceptions) to maintain, validate and append new transactions to the database. This decentralized capability is inherent in the coding of the blockchain (and is part of its governance model), and this allows for participants who do not need to trust or even know each other. Hence, public blockchain exists and operates as a trustless system.

[53] StackExchange, "Questions on the terms 'distributed' and 'decentralized,'" *https://ethereum.stackexchange.com/questions/7812/question-on-the-terms-distributed-and-decentralised*
[54] *Ibid.*

ISACA. All Rights Reserved.

Blockchain technology allows for the ordering of data in a list of blocks, which are usually a collection of transactions. Two specific structures are critical to the process of blocks being appended to the blockchain:[55]

- **Pointers**—variables that store information regarding the position of another data set or variable.
- **Linked lists**—a sequence of blocks, in which each block contains specific data sets and which are connected or linked to a subsequent block within the blockchain, with the direction of the pointer (**figure 4.3**).

Figure 4.3—Linked Lists

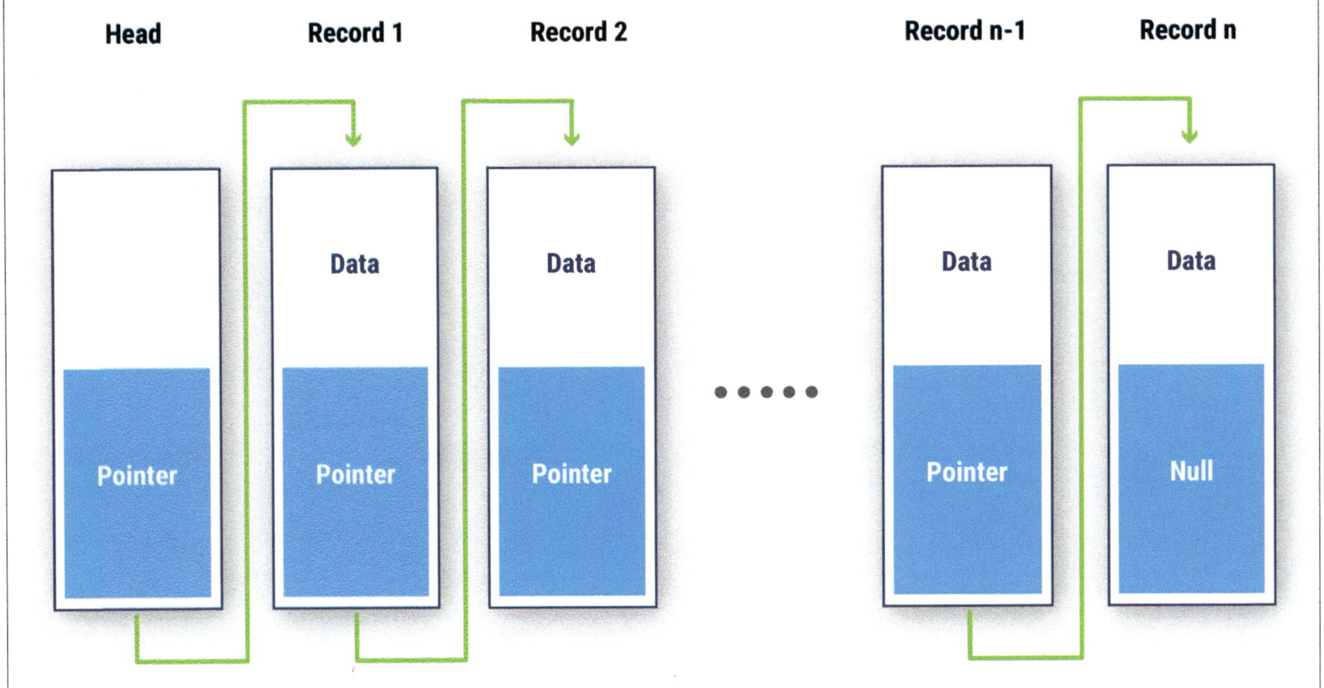

Source: Lastovetska, A.; "Blockchain Architecture Basics: Components, Structure, Benefits & Creation," MLSDev, 31 January 2019, *https://mlsdev.com/blog/156-how-to-build-your-own-blockchain-architecture*

Hashing is the functional process denoted by the pointer and linked lists method. For any given blockchain, the first block created or mined (known in common parlance as the genesis block) has no pointer to it at all, because there is no preceding block. **Figure 4.4** illustrates the pointer/linked-list paradigm in the context of a blockchain database. Note that the unique pointer or hash is derived from the previous block header. Most blockchain processes use a Merkle root, which is the hash of all prior hashes of transactions on the particular blockchain. The Merkle root is included in the subsequent block header. It is in this way that it becomes possible to securely verify that a transaction has been accepted by the network.[56]

[55] Lastovetska, A.; "Blockchain Architecture Basics: Components, Structure, Benefits & Creation," MLSDev, 31 January 2019, *https://mlsdev.com/blog/156-how-to-build-your-own-blockchain-architecture*

[56] StackExchange, "What is the Merkle root?," *https://bitcoin.stackexchange.com/questions/10479/what-is-the-merkle-root*

ISACA. All Rights Reserved.

Figure 4.4—Pointer/Linked List Paradigm

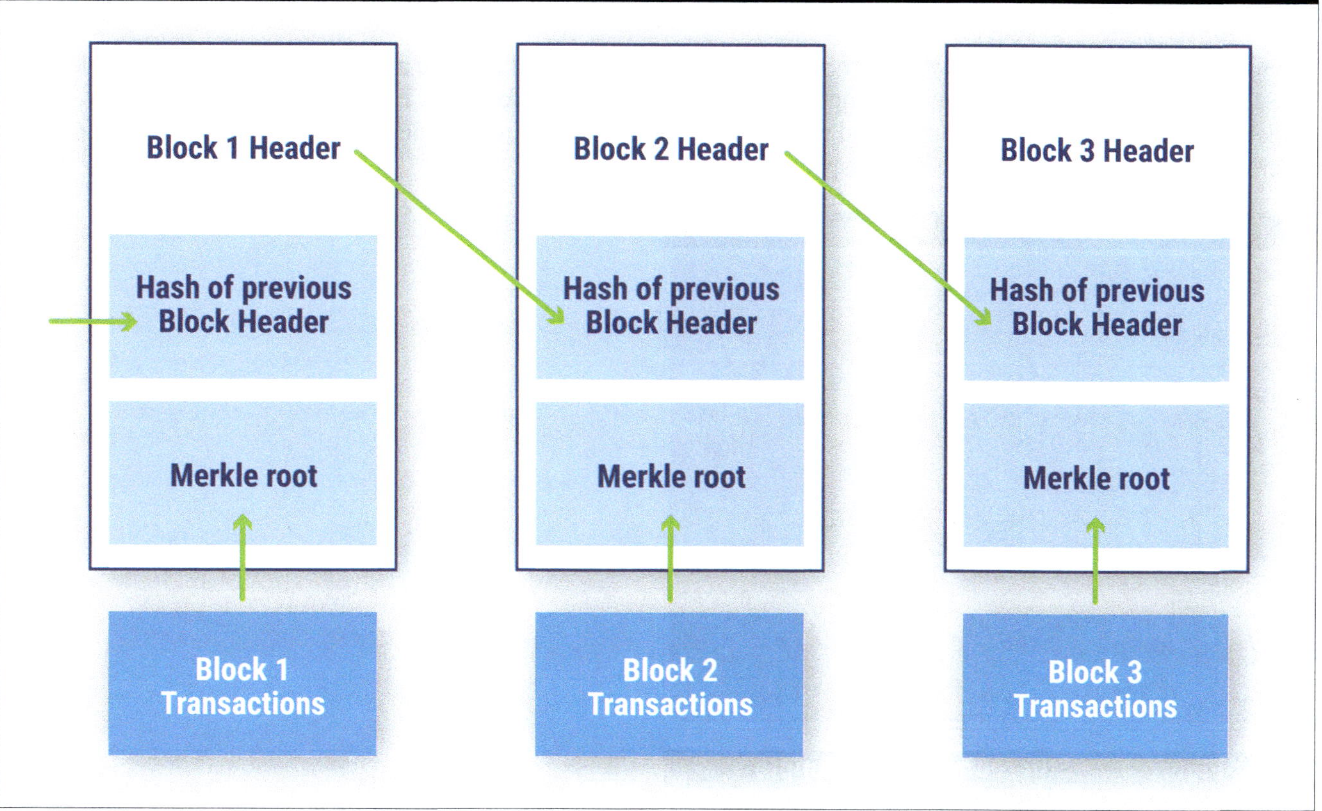

Source: Lastovetska, A.; "Blockchain Architecture Basics: Components, Structure, Benefits & Creation," MLSDev, 31 January 2019, *https://mlsdev.com/blog/156-how-to-build-your-own-blockchain-architecture*

To summarize and using the Bitcoin blockchain as an example, the usual components of a block are shown in **figure 4.5**.

Figure 4.5—Components of a Block

Software version	Only relevant if miners have differing software versions.
Hash of previous block	In seconds, Universal Time Coordinated (UTC)
Root hash of Merkle tree	All transactions contained in the block, aggregated into a hash
Goal of current difficulty level	Directly as a result of proof of work in the Bitcoin blockchain, the difficulty level indicates how small the new hash must be to claim validity. The lower the goal is, the harder it will be to find a matching hash.
Nonce	This is the variable incremented by the proof of work. The nonce is the first number that a blockchain miner must discover before solving for a block in the blockchain and receiving a reward.

Source: Vidrih, M.; "What is a Block in the Blockchain?," Medium, 29 December 2018, *https://medium.com/datadriveninvestor/what-is-a-block-in-the-blockchain-c7a420270373*

When a miner on a blockchain network works to add a new block of transactions to the blockchain, it broadcasts that block to all nodes on the network. Based on the validity of that blocks' data (i.e., the validity of the key signatures, and the hashed transaction[s] within the block), the network of nodes will accept or reject the block. The following content analyzes the different types of nodes, including miner, at a holistic level and uses the Bitcoin blockchain as

ISACA. All Rights Reserved.

an example. Note that different blockchain implementations, both public and private, have different miner/node/consensus methodologies, or in some instances, none at all.

The use of blockchain explorers, which are web-based tools to track and analyze a wide variety of public blockchain transactions, has grown with the wider adoption of blockchain and cryptoassets. The following figures from Tradeblock[57] capture detailed technical and monetary information for Bitcoin block number 123456 (**figure 4.6**) and Ethereum block 10304167 (**figure 4.7**), respectively.

Figure 4.6—Bitcoin Block Number Example

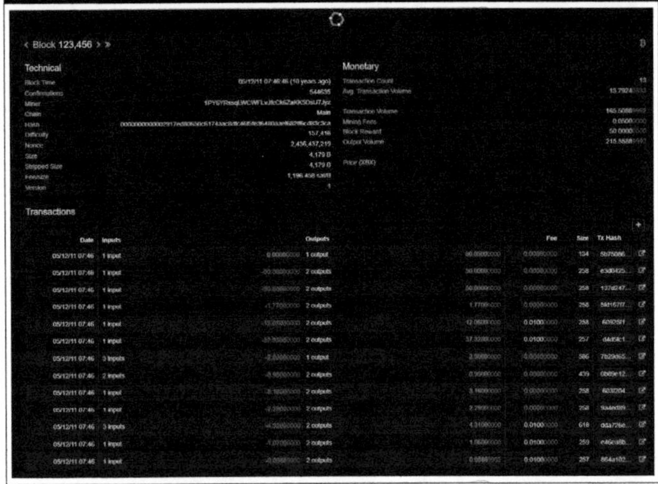

Figure 4.7—Ethereum Block Example

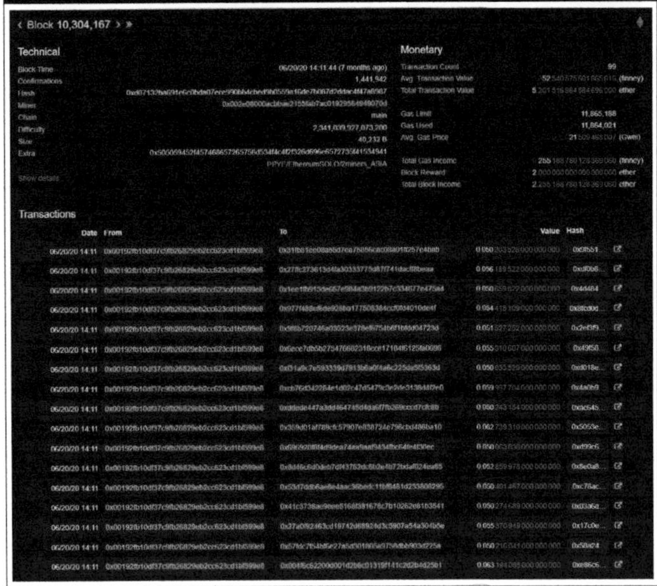

4.3.1 Blockchain Nodes

In standard IT nomenclature, a network node is a point at which a message or data packet can be created and/or transmitted. The Bitcoin blockchain has several node types.

[57] TradeBlock, *https://tradeblock.com/home*

ISACA. All Rights Reserved.

Full nodes

Bitcoin blockchain full nodes support and provide security for the Bitcoin blockchain and are critical to network functioning. They are also sometimes referred to as fully validating nodes, clearly noting their function of processing transactions into blocks in line with the network consensus methodology rules. Full nodes are able to relay new transactions and blocks to the blockchain. Usually (with very limited exceptions), a full node downloads the entire copy of the Bitcoin blockchain software and data set.

Miners

A miner or mining node is a node that contributes to the network by guessing the combinations needed to confirm the blocks of transactions to be added to a blockchain, producing new bitcoins in the process. The full node validates this information, confirms its veracity and publishes this information to the other nodes on the network. This is what allows a transaction of bitcoin between two parties to be considered completed. Mining nodes can be considered a subset of full nodes.

Listening nodes

Some blockchain networks allow for listening nodes, which are publicly visible full nodes. A listening node can communicate and share data or information to any other node that connects with it, and essentially acts as a sharing point for data in the network. Any full node that is fully publicly accessible is considered a listening node.

Masternodes

Selected blockchain implementations allow for masternodes. These nodes are more weighty than normal nodes, in that they can process all the functions of full nodes or miners but are also able to facilitate other processes in a blockchain, such as the enforcement of network voting rights or increasing privacy on certain transactions. Given their power, often masternodes are required to post a significant amount of collateral in the form of cryptocurrency as a protection against abuse of these abilities. In addition, they tend to be very computationally expensive. Dash,[58] as a fork of the Bitcoin blockchain, was the first blockchain network to use masternodes.

It is important to note that all nodes can be offline or online. If online, they are receiving, mining or broadcasting blocks to the network, in their applicable roles; whereas, offline nodes are not. When back online, offline nodes are required to catch up with the rest of the blockchain by downloading all data and blocks they missed since their sign-off from the network (otherwise known as synchronizing with the blockchain).

4.3.2 Blockchain Network Interoperability

Lately, one of the more important conversations in the global blockchain arena is that of blockchain interoperability. Blockchain interoperability is the "ability to share information across different blockchain networks, without restrictions."[59]

Given the multitude of different public and private blockchains currently in existence and the mass of blockchain use cases arising across different industries, it was only a matter of time before the importance of disparate blockchains being able to communicate and share data with each other became apparent. The World Economic Forum (WEF) offers an interesting definition of blockchain interoperability:

[58] Dash, *https://www.dash.org*
[59] O'Neal, S.; "Blockchain Interoperability, Explained," Cointelegraph, 5 September 2019, *https://cointelegraph.com/explained/blockchain-interoperability-explained*

ISACA. All Rights Reserved.

[I]nteroperability is: a) the capacity of computer systems to exchange and make use of information; and b) the capacity to transfer an asset between two or more systems while keeping the state and uniqueness of the asset consistent.[60]

In the course of ordinary network interaction, interoperability is a rather straightforward concept, but, in the world of blockchain, this capacity to transfer an asset makes blockchain interoperability significantly more complex.

Two types of interoperability—data exchange and digital asset exchange—are described in the following sections.

Data Exchange

As the name implies, this type of interoperability references the exchanges of data, as opposed to some type of digital asset, between one blockchain and another, or initiating a function (such as might be embodied in a smart contract) from one blockchain to another. The essence of this exchange is rooted in the ability of an occurrence on one blockchain to affect another blockchain. In technical terms, this type of blockchain-to-blockchain call (via application programming interface [API]) can encompass many functions. For example, a smart contract on blockchain A interacts with blockchain B to confirm that a transaction on blockchain B has been completed and validated. This confirmation allows smart contract on blockchain A to release the title to an asset, for instance. Data exchange interoperability allows blockchains to reflect state changes on each other. Another example in the shipping industry regards:

the changing ownership of a bill of lading (BoL) from a shipper on Ethereum to a consignee on Hyperledger. The BoL is a document of title. Currently, the BoL is typically issued from the ocean carrier to the shipper/seller as proof of receipt of the container and contains data on... [the shipment]....[61]

In this example, the document can be exchanged multiple times between multiple parties. Interoperable blockchains allow this exchange to be a much more automated, efficient and cost-effective process.

Digital Asset Exchange

Digital asset exchange is, in some ways, a more thorough implementation of the interoperable blockchain model, in that it is designed to allow for the ability to exchange or transfer blockchain resident assets from one blockchain ecosystem to another. One example is the exchange or transfer of a cryptocurrency from one blockchain to another. On a technical level, much work is being done to enable this capability. The creation of code that lives as a programming layer above multiple blockchains and that aids in signature verification for transacting parties between these blockchains is one example.

There are only three main technical approaches to blockchain interoperability currently, although that may change in the future. Many of these approaches are addressed in the Vitalik Buterin paper[62] on blockchain interoperability. (Vitalik is one of the founders of the Ethereum protocol and blockchain.) These technical methods and some prominent examples are described in the following sections.

Hashed Timelocks

Primarily applicable to the exchange of digital assets, this technical approach is a type of smart contract used in cryptoasset transactions and is designed to remove counterparty risk, i.e., the risk that the other party to a transaction

[60] World Economic Forum, "Inclusive Deployment of Blockchain for Supply Chains: Part 6 – A Framework for Blockchain Interoperability," 9 April 2020, *https://www.weforum.org/whitepapers/inclusive-deployment-of-blockchain-for-supply-chains-part-6-a-framework-for-blockchain-interoperability*

[61] *Ibid.*

[62] Larsen, A.; "A Primer on Blockchain Interoperability," Blockchain Capital, 9 December 2018, *https://blockchain.capital/top-highlight-a-primer-on-blockchain-interoperability/*

ISACA. All Rights Reserved.

cannot participate in the trade. It allows for a transaction to be bound to a time range specified in the smart contract. An example is a seller of a cryptoasset on one blockchain has to confirm receipt of payment from the buyer on another blockchain, in a way that is cryptographically proven, i.e., hashed. If this is not accomplished, the transaction is rejected and cannot occur. Atomic swaps are a prominent blockchain solution that uses hashed timelocks.

Atomic Swaps

Atomic swap is a peer-to-peer exchange of cryptoassets from one party to another, who are on two separate blockchains, without needing a third-party intermediary like a broker or exchange. Throughout an atomic swap operation, both parties maintain complete ownership of and control over their private keys. **Figure 4.8** provides a graphical representation of atomic swaps.

Figure 4.8—Atomic Swaps

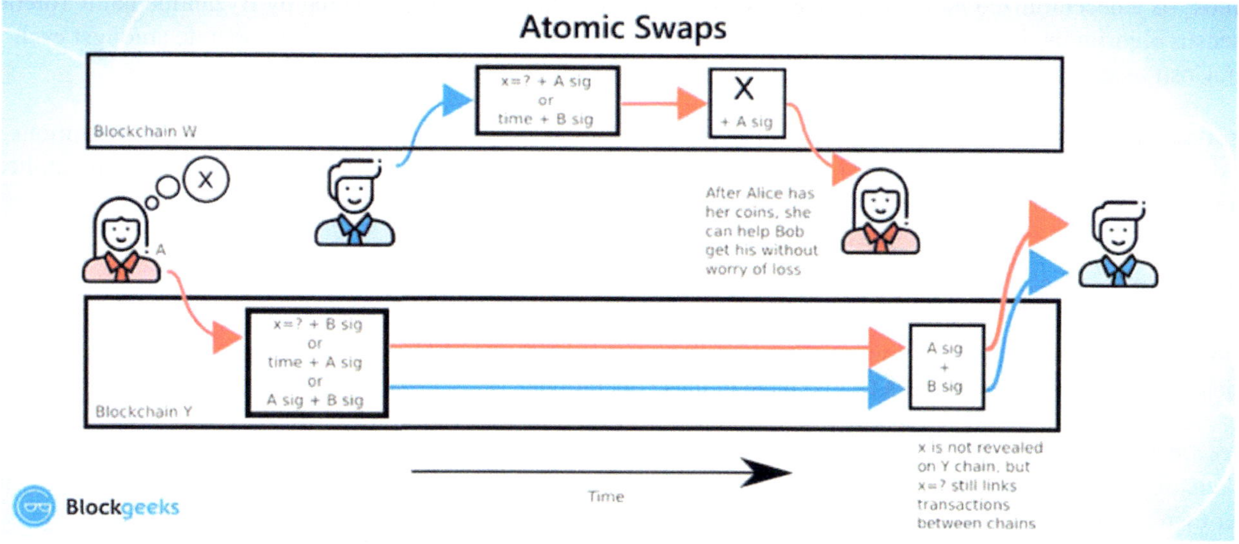

Source: Uczciwek, M.; "What Are Atomic Swaps? The Most Comprehensive Guide Ever!," Blockgeeks.com, *https://blockgeeks.com/guides/atomic-swaps/*

Sidechains

A sidechain is a blockchain that works alongside a primary blockchain, so that data entries or transactions from primary blockchain can be linked or operate both from and to the sidechain. Some sidechain implementations can operate independent to their primary. Some sidechains are connected to the parent blockchain using a two-way peg, which allows for the exchange of assets at a predetermined rate between both. Many sidechains use the hashed timelocks capability.

Notary Schemes

Relevant primarily for data exchange between blockchains, notary schemes leverage a trusted coalition or federation to confirm actions that have occurred on another blockchain. For example, the coalition can verify to blockchain A that an event on blockchain B occurred, or that some fact or data point about blockchain B is true.

The coalition uses a consensus algorithm to issue a cryptographic signature that is used to finalize the transaction. In this case, a separate blockchain ledger always exists, and the coalition manages it to ensure interoperability between the two other blockchains. Implementation-wise notary schemes are generally easier but require all parts to accept a

ISACA. All Rights Reserved.

control by the coalition. It cannot be a trustless environment. The atomic mode of the Interledger Protocol[63] uses the notary model.

Relays

Primarily applicable to data exchange between blockchains, relays allow blockchains themselves to verify transactions or events that have occurred on other blockchains. For example, relay is a smart contract on blockchain A that operates as a light client of blockchain B, engaging with blockchain B's consensus methods to verify block headers. This gives blockchain A the capability to recognize and confirm specific event changes on blockchain B. Thus, relays allow blockchains to verify events that have taken place on other blockchains on a chain-to-chain basis, without the requirement for a trusted node of a notary. Cosmos is a prominent example of an interoperability solution using relays.

Cosmos[64] is a decentralized network of independent parallel blockchains, each powered by Byzantine Fault Tolerant consensus algorithms. It is designed as an ecosystem of blockchains that can scale and interoperate amongst each other. Cosmos uses relayers to post specific event information to the relevant smart contract.

Several methods have evolved to enable interoperability. Although none of these cross-chain interaction solutions are in wide-scale usage, it is clear that one of the roadblocks to wider enterprise adoption is the lack of interoperability, and mass adoption will be hindered until resolved.

Blockchain Nodes and APIs

It is important to briefly address how network communications between blockchain nodes operate, and the role that application programming interfaces (APIs) have in these operations.

Nodes are a critical part of a blockchain ecosystem. In a decentralized peer-to-peer system (at least for public blockchains), all nodes act as a combined client and server. Thus, the responsibilities of nodes are protocol-specific rather than software-specific. In communicating with each other, blockchain nodes use common protocols. For example, the nodes on the Bitcoin blockchain randomly connect to other nodes and use simple broadcasting to propagate transactions and blocks over TCP.[65]

Commonly, APIs act as computing interfaces to define the interactions between various software intermediaries. They define the kinds of calls or requests that can be made, how to make them, data formats and process conventions that should be followed.[66] Given their functions, blockchains can have a variety of APIs, including wallet APIs, blockchain information APIs and WebSocket APIs for streaming data. In addition, a wide variety of blockchain API providers[67] currently exist that offer ready-made code to initiate blockchain data queries. Some common examples include:

- Plain text query to retrieve data from blockchain, which offers a range of query types, including real-time data, address lookups and transaction lookups.[68] Following is an example is this simple address lookup from Blockchain.com.

[63] Interledger, *https://interledger.org*
[64] Cosmos, *https://cosmos.network*
[65] Bitcoin Wiki, "Network," *https://en.bitcoin.it/wiki/Network*
[66] Wyatt, M., "What is an API? A Digestible Definition with API Examples for Ecommerce Owners," BigCommerce, *www.bigcommerce.com/blog/what-is-an-api/#conclusion*
[67] Yusov, K.; "How to Choose the Best Blockchain API for Your Project," Jelvix, *https://jelvix.com/blog/how-to-choose-the-best-blockchain-api-for-your-project*
[68] Blockchain, "Query API: Plaintext query api to retreive data from blockchain.info," *https://www.blockchain.com/api/q*

ISACA. All Rights Reserved.

Example:

Address Lookups

To filter by x number of confirmations include the confirmation parameter, e.g.:

/q/addressbalance/1EwoHtiXB4iFwedPr49iywjZn2nnekhoj?confirmations=6

- Wallet APIs that allow for the creation and the sending and receiving of payments between blockchain wallets,[69] available in Python,[70] Java, .NET[71] and Ruby.[72]
- Streaming socket with low latency, providing data on blocks and transactions, including subscribing to unconfirmed transactions on the blockchain, activity for a specific wallet address and notifications for when new blocks are found and added to a blockchain.[73]

4.3.3 Blockchain Specific Hardware

Since the launch of the Bitcoin blockchain in 2009, there has been an increase in the number of public and private blockchains. The growing number of public blockchain implementations requires, for the most part, a commensurate growth in not just the number of consensus protocols, but also the computational effort required to fulfill these protocols and operate these blockchains. The more the network grows, the harder the problems become to solve and the more computationally expensive it becomes to be a miner on a public blockchain.

Given this growing computational cost and the incentive mechanisms of consensus methodologies, such as proof of work, a number of technological solutions have been developed to allow blockchain nodes and miners to successfully mine new blocks and receive rewards, such as new bitcoin.

In particular, over the past several years specific chips have been designed or repurposed to enable faster blockchain processing. Specifically, there has been significant growth in the use of graphics processing units (GPU) and application-specific integrated circuits (ASIC), to allow for blockchains to operate faster and more efficiently, thus enabling greater blockchain mining output.

The use of GPUs for blockchain came to the fore relatively quickly. In the early days of the Bitcoin blockchain, using the central processing unit (CPU) of a home computer was enough to facilitate someone becoming a bitcoin miner. Over time and with the growth of mining complexity, it became clear that GPUs, which are generally associated with the video rendering systems of a computer,[74] were more efficient and faster than CPUs. A sample CPU core can execute four 32-bit instructions per clock. A GPU can potentially execute 3200 32-bit instructions per clock, a difference of 800 times more instructions per clock.[75] The difference in computational speed, as well as the growing ability to be rewarded with bitcoin for being a winning miner, becomes clear when reviewing these numbers.

As this computational arms race continued to heat up, even the use of GPUs proved too little to meet growing computational demands. Around 2017, bitcoin mining reached what could be considered industrial proportions, with miners joining together to create entire server farms with thousands of GPU cards and peta hash level capacities to mine bitcoin and other cryptocurrencies. Indeed, chip manufacturers had to ramp up production as video gamers encountered GPU shortages because of blockchain miners.[76]

[69] Blockchain, "Blockchain Wallet API: Simple API for Blockchain Wallet users to send and receive bitcoin payments," *https://www.blockchain.com/api/blockchain_wallet_api*
[70] Python, *https://www.python.org*
[71] Microsoft, *https://dotnet.microsoft.com/learn/dotnet/what-is-dotnet*
[72] Ruby, *https://www.ruby-lang.org/en/*
[73] Blockchain, "WebSocket API: Real-Time blockchain data," *https://www.blockchain.com/api/api_websocket*
[74] Bitcoin Wiki, "Why a GPU mines faster than a CPU," *https://en.bitcoin.it/wiki/Why_a_GPU_mines_faster_than_a_CPU*
[75] *Ibid.*
[76] Magas, J.; "Top Five Biggest Crypto Mining Areas: Which Farms Are Pushing Forward the New Gold Rush?" Cointelegraph, 23 June 2018, *https://cointelegraph.com/news/top-five-biggest-crypto-mining-areas-which-farms-are-pushing-forward-the-new-gold-rush*

ISACA. All Rights Reserved.

However, even the use of GPUs proved to be insufficient over time to meet the need for processing speed and power in the cryptocurrency world. Thus, ASICs were used to further the goals of miners. Unlike CPUs or GPUs, ASIC devices are designed and built specifically for the sole purpose of mining. For the most part, each ASIC miner is created to mine a specific cryptocurrency.[77] Developing and manufacturing ASICs as mining devices is not only complex but relatively costly, and they are more difficult to acquire than GPUs. However, they engage in the mining activity of a blockchain significantly faster than less powerful computers.

Comparatively speaking, if GPUs are faster than CPUs, then ASIC is faster than GPU (**figure 4.9**). ASIC is much faster. For some perspective regarding ASIC power, a Bitcoin-specific ASIC can calculate hashes 100,000 times faster than even the more powerful CPU.[78] Concerning is that large concentrations of ASIC miners can lead back to centralization. For example, an issue arose in 2019 in the case of Bitcoin in which for a period of time, the majority of miners are located in China.[79]

Figure 4.9—Computational Power for Processing Speed and Power

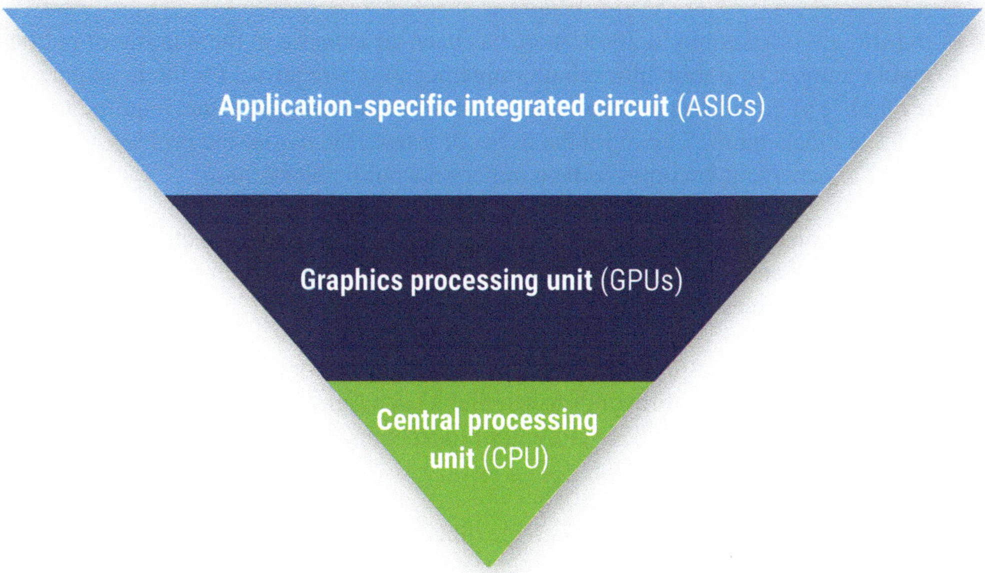

This section only briefly discusses the growing hardware challenges associated with blockchain. From the use of specialized chips, to the growth of mining pools and the latest chip designs, it is clear that cryptocurrencies and blockchain are innovative in hardware creation and usage.

[77] Tardi, C.; "Application Specific Integrated Circuit (ASIC) Bitcoin Miner," Investopedia, 2 September 2019, *https://www.investopedia.com/terms/a/asic.asp*

[78] Compute North, "What is the difference between ASIC and GPU cryptocurrency mining?," 1 May 2018, *www.computenorth.com/what-is-the-difference-between-asic-and-gpu-cryptocurrency-mining/*

[79] CoinShares Research, "The Bitcoin Mining Network: Trends, Average Creation Costs, Electricity Consumption & Sources," December 2019, *https://coinshares.com/research/bitcoin-mining-network-december-2019*

ISACA. All Rights Reserved.

4.4 Chapter 4 Knowledge Check

REVIEW QUESTIONS

1. An example of a centralized infrastructure topology is:

   A. Client-server
   B. Peer-to-peer
   C. Cloud computing
   D. Mesh network

2. ________________ nodes on a blockchain network are also known as fully validating nodes.

   A. Miner
   B. Listening
   C. Validating
   D. Full

3. A subset of full nodes on a blockchain network are ________ nodes.

   A. Miner
   B. Listening
   C. Validating
   D. Full

4. Blockchain interoperability can be defined as:

   A. Creating a standard for blockchain hardware
   B. The ability to share data across blockchain networks
   C. Standardizing blockchain-mining processes
   D. The proprietarization of data APIs on a blockchain system

5. A peer-to-peer exchange of a cryptoassets between blockchain systems without need for a third party is known as a:

   A. Sidechain
   B. Relay
   C. Atomic swap

6. A secondary chain that can be linked to data in the primary blockchain is known as a:

   A. Sidechain
   B. Relay
   C. Atomic swap
   D. Digital asset exchange

7. _________________ enable interactions among various software intermediaries.

   A. TCP communications
   B. Peer-to-peer systems
   C. APIs
   D. WebSocket

ISACA. All Rights Reserved.

8. ASIC stands for:

A. Analysis system for index counters
B. American standard of interoperability circuits
C. Application-specific integrated circuits
D. Application systems of interoperability counters

Answers on page 63

ISACA. All Rights Reserved.

Chapter 4 ANSWER KEY

Review Questions

1. **A. Client-server. Refer to page 51.**
   B. Peer-to-peer
   C. Cloud computing
   D. Mesh network

2. A. Miner
   B. Listening
   C. Validating
   D. Full. Refer to page 55.

3. **A. Miner. Refer to page 55.**
   B. Listening
   C. Validating
   D. Full

4. A. Creating a standard for blockchain hardware
   B. The ability to share data across blockchain networks. Refer to page 55.
   C. Standardizing blockchain-mining processes
   D. The proprietarization of data APIs on a blockchain system

5. A. Sidechain
   B. Relay
   C. Atomic swap. Refer to page 57.

6. **A. Sidechain. Refer to page 57.**
   B. Relay
   C. Atomic swap
   D. Digital asset exchange

7. A. TCP communications
   B. Peer-to-peer systems
   C. APIs. Refer to page 58.
   D. WebSocket

ISACA. All Rights Reserved.

8. A. Analysis system for index counters
   B. American standard of interoperability circuits
   C. Application-specific integrated circuits. Refer to 59.
   D. Application systems of interoperability counters

ISACA. All Rights Reserved.

BLOCKCHAIN
FUNDAMENTALS
An ISACA Certificate

Chapter 5:

Business Usage and Considerations

Business Usage and Considerations

5.1 Learning Objectives 66
5.2 Overview 66
5.3 Major Blockchain Platforms 66
5.4 Example Use Cases 67
5.5 Healthcare 69
5.6 Other Industries 70
5.7 Tokenization and Digital Assets 70
5.8 Token Technology Standards 71
5.9 Blockchain Validity and Audit Considerations 73
5.10 Conclusion 74
5.11 Chapter 5 Knowledge Check 75

ISACA. All Rights Reserved.

Business Usage and Considerations

5.1 Learning Objectives

After completing this chapter, learners will be able to:

- Identify major blockchain platform providers and services.
- Describe example uses for blockchain technology.
- Articulate the importance of blockchain and digitization of assets.
- Distinguish among different token types and their uses in a blockchain system.
- Explain considerations for blockchain system audit and validity.

5.2 Overview

Demand is growing to use blockchain technology across an ever widening number of business enterprises and industries. It is estimated that blockchain technology could generate economic value of more than $300 billion by 2027, across a multitude of sectors.[80] This section analyzes some major blockchain platforms, and current and future use cases.

5.3 Major Blockchain Platforms

5.3.1 Amazon Web Services (AWS)

Amazon, the American multinational conglomerate and technology company, entered the blockchain arena with the launch of Amazon Managed Blockchain.[81] Amazon Managed Blockchain is a fully managed service that makes it relatively easy to create and manage scalable blockchain networks using Hyperledger Fabric and Ethereum. The solution enables participants to share data and execute transactions without a central authority, and by leveraging the AWS ecosystem, it eliminates the need to manually provision hardware, configure software or set up networking and security components.

5.3.2 Ethereum

In addition to being one of the largest cryptocurrency platforms by market capitalization, Ethereum[82] is a decentralized open-source blockchain featuring smart contract functionality. Ether is the cryptocurrency generated by Ethereum miners as a reward for computations performed to secure the blockchain. By May of 2020, Ethereum served as the platform for over 260,000 different cryptocurrencies, including four of the top 10 cryptocurrencies by market capitalization.

Ethereum offers a decentralized virtual machine, the Ethereum Virtual Machine (EVM), that can execute scripts using an international network of public nodes.

[80] Fintechnews Switzerland, "Fintech Revenues to Reach US$500B by 2030: UBS Research," 7 May 2020, *https://fintechnews.ch/fintech/fintech-revenues-to-reach-us500b-by-2030-ubs-research/35500/*
[81] Amazon Web Services, "Amazon Managed Blockchain" *https://aws.amazon.com/managed-blockchain/*
[82] Wikipedia, "Ethereum," *https://en.wikipedia.org/wiki/Ethereum*

ISACA. All Rights Reserved.

5.3.3 Hyperledger

Launched in December 2015, Hyperledger[83] is an umbrella project of open-source blockchains and related tools, and was started by the Linux Foundation, with participation by IBM, Intel and SAP. The goal of Hyperledger is to support the collaborative development of blockchain-based distributed ledgers and advance cross-industry collaboration, with a particular focus on improving performance and reliability of blockchain-based systems. Notable Hyperledger frameworks include: Fabric, Iroha, Sawtooth and Besu. Their toolsets (Caliper, Cello, Explorer, Quilt and Ursa) offer a range of tools for querying, collaboration, deployment and more.

5.3.4 Microsoft Azure

Microsoft Azure[84] Blockchain Service is a fully managed ledger service that enables users the ability to grow and operate blockchain networks at scale in Azure. By providing unified control for both infrastructure management and blockchain network governance, Azure Blockchain Service provides simple network deployment and operations, built-in consortium management, and smart contract development. The service is designed to support multiple ledger protocols, and, currently, it provides support for the Quorum ledger (itself based upon Ethereum).

5.3.5 R3

R3[85] is an enterprise blockchain technology company and consortium with a network of more than 300 firms collaborating to build distributed applications utilizing Corda, the R3 blockchain technology that records, manages and synchronizes financial agreements, and standardizes data and business processes. R3 focuses on blockchain adoption across several industries, including financial services, insurance, healthcare, trade finance and digital assets.

5.4 Example Use Cases

It is important to point out that the adoption of blockchain, although occurring in fits and starts, is proceeding apace across multiple industries and use cases. This section reviews the most relevant use cases for three types of industries: financial markets, supply chain and healthcare. It is important to be aware of the cybersecurity protocols and compliance requirements within these industries. For example, financial markets are subject to Bank Secrecy Act (BSA)[86] and Anti-Money Laundering (AML)[87] rules, among many others globally; healthcare in the United States is subject to Health Insurance Portability & Accountability Act (HIPAA)[88] and personally identifiable information is protected in Europe under General Data Protection Regulation (GDPR).[89]

5.4.1 Financial Markets

Broadly speaking, financial markets were the first to embrace the potential uses and benefits of blockchain technology. Those in the financial industry and blockchain technologists find a certain irony in this, considering that substantial numbers of global financial markets participants and business exist and make their living as intermediaries, which is what blockchain is designed to eliminate. Despite that irony, much progress has been made in allowing for blockchain technologies to become integrated into financial markets. One prominent example is clearing and settlement.

[83] Wikipedia, "Hyperledger," *https://en.wikipedia.org/wiki/Hyperledger*
[84] Microsoft, "What is Azure Blockchain Service?," *https://docs.microsoft.com/en-us/azure/blockchain/service/overview*
[85] R3, *https://www.r3.com*
[86] Office of the Comptroller of the Currency, "Bank Secrecy Act," *www.occ.treas.gov/topics/supervision-and-examination/bsa/index-bsa.html*
[87] FINRA, "Anti-Money Laundering (AML)," *www.finra.org/rules-guidance/key-topics/aml*
[88] California Department of Health Care Services, "Health Insurance Portability & Accountability Act," *www.dhcs.ca.gov/formsandpubs/laws/hipaa/Pages/1.00WhatisHIPAA.aspx*
[89] GDPR EU, "What is GDPR, the EU's new data protection law?," *https://gdpr.eu/what-is-gdpr/*

ISACA. All Rights Reserved.

Clearing and Settlement

Clearing and settlement jointly are the processes by which financial trades settle—that is, the timely and accurate transfer of funds to a seller and securities to a buyer. Usually, a dedicated organization acts as an intermediary and assumes the role of buyer and seller for transactions, to reconcile trades between transacting parties and match buy and sell orders in the market.[90]

Settlement specifically is the process by which securities or interests in securities are delivered, usually against monetary payment of money, to fulfill contractual obligations from a trade. As early as 2015, the Australian Stock Exchange (ASX), the worlds sixteenth-largest exchange,[91] sought to replace its legacy CHESS clearing and settlement system with a blockchain-based platform. In early 2016, it selected Digital Asset Holdings, a blockchain and financial technology provider,[92] as a technology partner to develop, test and demonstrate a working blockchain prototype of an equities post-trade platform. Like many other major technology projects of this scope, the ASX recently announced that the project go-live date will move from April 2021 to an as-yet unspecified date in the future,[93] but work continues with the ASX and other exchanges around the world to leverage the efficiencies and potential cost benefits of blockchain technology.

5.4.2 Supply Chain

Another prominent industry seeking to benefit from the power of blockchain technology is the supply chain industry. Supply chain refers to the system of organizations, activities, people, information and resources involved in supplying a product or service to a consumer.[94] Supply chain activities involve the change of natural resources, raw materials and components into a finished product that is delivered to the end customer, and all the logistical work associated with delivering these products or services. Blockchain technology has made substantial inroads into the supply chain ecosystem, as indicated in the following examples.

Walmart and the IBM Food Trust

Given the complexities of the global food supply chain, it is not surprising that one of the world's largest food shippers would consider blockchain technology for its vast supply chain ecosystem. For example, a typical product recall costs $8-million per year, on average. The use of blockchain can potentially reduce this with:

- Improved audit features and cost savings of blockchain technology.
- Ability to securely and trustlessly track the provenance of goods allows for tighter controls on inventory and less product leakage or spoilage.
- Streamlining of insurance processes, better inventory control, more effective management and auditor analysis, and overall better forecasts across the entire supply chain and its members.

Thus, in 2018 and in the aftermath of an e-coli outbreak of its food produce, Walmart, the American multinational retail corporation, began working with IBM to employ blockchain to digitize the food supply chain process. Their effort led to the launch of the IBM Food Trust,[95] which leverages Hyperledger Fabric to reinvent the food supply chain from farm to producer to delivery.

[90] Chen, J.; "Clearing," Investopedia, 9 April 2019, *www.investopedia.com/terms/c/clearing.asp*
[91] Wikipedia, "Australian Security Exchange," *https://en.wikipedia.org/wiki/Australian_Securities_Exchange*
[92] Digital Asset, *https://digitalasset.com*
[93] Finextra, "ASX rips up go-live timetable for Chess replacement system," 25 March 2020, *www.finextra.com/newsarticle/35516/asx-rips-up-go-live-timetable-for-chess-replacement-system*
[94] Lutkevich, B.; "supply chain," TechTarget, *whatis.techtarget.com/definition/supply-chain*
[95] *Op cit* IBM Food Trust

ISACA. All Rights Reserved.

Previously, it was reported that it typically took approximately seven days to trace the source of food. The latest platform using blockchain technology has demonstrably reduced that to 2.2 seconds,[96] thereby substantially reducing the possibility that contaminated food can reach wholesale and retail customers. This pilot program has been so successful that Walmart now traces over 25 products from five suppliers using this blockchain platform, with ultimate plans that all vendors to Walmart much be a part of this platform. Other major producers and shippers, such as Kroger and Nestle, have followed suit.

5.5 Healthcare

Healthcare has been one of the later, although no less important, entrants into the group of industries exploring the benefits of blockchain technology. For example, securing patient medical data, the tracing of counterfeit drugs or the secure management and tracking of prescription information are all important areas of research and prototype development in this space. Given the fact that the global health care industry is anticipated to reach almost $12 trillion by the year 2022,[97] the importance and potential cost savings that can come from using blockchain technology becomes ever more apparent. The following sections examine the securing of patient medical.

5.5.1 BurstIQ and Patient Records

Keeping patient medical data safe and secure, in an effort to comply with existing regulations and protect patient data from theft, is one of the preeminent areas of research and investment for blockchain technology. Security of data is clearly on ongoing and growing concern. Between 2009 and 2017, more than 170-million patient records were exposed to data breaches, resulting in the theft of, not just banking and credit card information, but also health and genomic testing records.[98] The ability of blockchain technology to keep a decentralized, incorruptible record of all patient data is one clear solution. Coupled with the ability of a blockchain implementation to mask specific personally identifiable information (PII), it becomes, in the eyes of many industry practitioners, an ideal platform to secure sensitive medical data in a reliable and legally compliant way.

From an ecosystem standpoint, one can envision a solution that allows hospitals, laboratories, doctors and patients to all be participants in a blockchain-based solution, with distinct governance models and defined rights to read, add or access specific information on the blockchain. BurstIQ offers a:

> *first-of-its-kind data management platform that builds multi-dimensional profiles of people, places and things, and enables the connections between them. It allows data to be securely owned, shared, and trusted across a global data exchange network, driving deeper insights and more meaningful collaborations in health.*[99]

With a blockchain platform designed specifically for the healthcare industry, BurstIQ claims to process over 50-billion data points, allowing for capabilities, such as the granular ownership of patient data, secure consent protocols, governance that cryptographically enforces ownership, secure on-chain data management, and multiple healthcare data profiles.[100] There are several other technology suppliers working to evolve the healthcare industry,[101] with the value of blockchain technology in the global healthcare industry expected to exceed $1.6 billion by 2025.[102]

[96] Miller, R.; "Walmart is betting on the blockchain to improve food safety," Techcrunch, 24 September 2018, *https://techcrunch.com/2018/09/24/walmart-is-betting-on-the-blockchain-to-improve-food-safety/*

[97] Business Wire, "The $11.9 Trillion Global Healthcare Market: Key Opportunities & Strategies (2014-2022) – ResearchAndMarkets.com," 25 June 2019, *www.businesswire.com/news/home/20190625005862/en/11.9-Trillion-Global-Healthcare-Market-Key-Opportunities*

[98] Daley, S.; "15 Examples of How Blockchain is Reviving Healthcare," Built In, 24 March 2020, *https://builtin.com/blockchain/blockchain-healthcare-applications-companies*

[99] Gallagher, A.; "Meet the Startups Using the BurstIQ Platform to Solve COVID-19," BurstIQ, *www.burstiq.com/meet-the-startups-using-the-burstiq-platform-to-solve-covid-19/*

[100] *Ibid.*

[101] Tinianow, A.; "Blockchain Technology is Already Improving Lives at 22 Hospitals," Forbes, 23 September 2019, *www.forbes.com/sites/andreatinianow/2019/09/23/blockchain-technology-is-already-improving-lives-at-22-hospitals/*

[102] Ugalmugle, S.; R. Swain; "Blockchain Technology in Healthcare Market Size: By Application (Data Exchange and Interoperability, Claims Adjudication and Billing Management, Drug Supply Chain Integrity, Clinical Trials, Cyber Security and Internet of Medical Things), By End-use (Pharmaceutical Companies, Healthcare Payers, Healthcare Providers), Industry Analysis Report, Regional Outlook, Application Potential, Price Trends, Competitive Market Share & Forecast, 2019 – 2025," Global March Insights, September 2019, *www.gminsights.com/industry-analysis/blockchain-technology-in-healthcare-market*

ISACA. All Rights Reserved.

5.6 Other Industries

The major blockchain platforms and multi-industry use cases represent only a fraction of the activity that is happening in these ecosystems. Rapid development and change can be expected in not only the industries listed above, but also across industries and ecosystems as diverse as digital media,[103] governmental records,[104] real estate,[105] transportation systems,[106] construction[107] and much more.

5.7 Tokenization and Digital Assets

This section discusses the issuance and usage of tokens, the tokenization process, and the importance of this emerging class of digital assets across multiple industries. Tokenization of assets, aside from the issuance and transacting of cryptoassets such as bitcoin, is the issuance of blockchain-based tokens representing real or hard assets.

The idea of tokenization goes back almost two decades, when the ability to use a token in digital commerce grew out of a desire to replace sensitive and private information with a nonsensitive digital equivalent. A solution was first introduced in 2001 by a company named TrustCommerce, a technology provider of secure, payment card industry (PCI) compliant payment processing solutions.[108] It was designed as a way to safely and securely store personal credit card information. Previously, merchants stored credit card data on their own servers, which clearly represents a security risk because anyone with access to the system was able to copy sensitive information. The TrustCommerce solution replaced a client account number with a randomized number, which they called a token. In processing payments, merchants referred to the token number only and TrustCommerce processed the payment on their behalf, eliminating the need for merchants to store any credit card or private data.[109]

Combining this concept of securing sensitive data using tokens with the power of blockchain technology represents a potent new evolution in data and asset exchange. Specifically, if a token is issued on a blockchain, it means that the blockchain records not just the issuance of the token, but also preserves a ledger of every single transaction of that token.

An important feature to consider is that in the same way blockchain solves the problem of double spending of cryptocurrencies, that feature can be assured for assets beyond solely digital or cryptocurrencies. Thus, in the same way that the Bitcoin blockchain is the decentralized, distributed ledger maintaining an immutable record of all Bitcoin transactions (and preventing double spending in the process), the tokenization process and blockchain can be used to represent the exchange of other assets.

Because this is a fast-evolving technology and ecosystem, there have been many attempts to define tokenization. Based on current industry convention, tokenization is the issuance of a cryptographically secured digital token on a blockchain, in which the token represents the full rights and ownership of an underlying tangible or intangible asset. Thus, economic value of the asset is conferred to and by the token, and ownership of the token represents ownership of the asset.

[103] IBM, "Now playing: transparency in media, entertainment and advertising with IBM Blockchain," *www.ibm.com/blockchain/industries/advertising-media*
[104] National Archives and Records Administration, "Blockchain White Paper," February 2019, *www.archives.gov/files/records-mgmt/policy/nara-blockchain-whitepaper.pdf*
[105] Liebkind, J.; "How Blockchain Technology is Changing Real Estate," Investopedia, 22 March 2020, *www.investopedia.com/news/how-blockchain-technology-changing-real-estate/*
[106] Astarita, V.; V.P. Giofre; G. Mirabelli; V. Solina; "A Review of Blockchain-Based Systems in Transportation," MDPI Open Access Journals, 29 December 2019, *www.mdpi.com/2078-2489/11/1/21/htm*
[107] Tapscott, D.; R.V. Vargas; "How Blockchain will Change Construction," *Harvard Business Review*, 26 July 2019, *https://hbr.org/2019/07/how-blockchain-will-change-construction*
[108] TrustCommerce, *www.trustcommerce.com*
[109] CoreLedger, "What is Tokenization? Everything You Should Know," Medium, 6 June 2019, *https://medium.com/coreledger/what-is-tokenization-everything-you-should-know-1b2403a50f0e*

ISACA. All Rights Reserved.

5.8 Token Technology Standards

5.8.1 Native Tokens

Native tokens exist at the foundational protocol level of a blockchain, are created at the genesis block and are unique to that blockchain. These tokens are core to the intrinsic functioning of their blockchain. They are usually used to reward the successful processing of transactions and/or the creation of blockchain blocks (such as in the Bitcoin blockchain). They also serve as the unit of account for their blockchain.

5.8.2 Layer 2 Tokens

Layer 2 tokens refer to a secondary coding that exists on top of the original blockchain coding structure. Secondary coding allows for the evolution of a decentralized blockchain to address limitations, including the scaling of speed and volume, the ability to create and execute smart contracts, and the ability to perform offchain transactions. Although there are many Layer 2 tokens, the Bitcoin Lightening Network is a prominent example. Bitcoin Lightning is a second-layer technology for bitcoin that uses micropayment channels to scale its blockchain capability to conduct transactions. By taking transactions away from the main blockchain, it is expected to decongest Bitcoin and reduce associated transaction fees.[110] Given its growing use in smart contracts and applications, the Ethereum blockchain has multiple standards (known as Ethereum Request for Comments [ERCs]) designed to enable Layer 2 tokens. Two of the most common are ERC-20 and ERC-721.

ERC-20

ERC-20 is the most common Ethereum token standard and the technical standard used for all smart contracts on the Ethereum blockchain for token implementation.[111]

ERC-721

ERC-721 is an open standard that describes how to build nonfungible or unique tokens on the Ethereum blockchain. Whereas, most tokens are fungible (every token is the same as every other token), all ERC-721 tokens are unique.[112]

5.8.3 Token Types

This section discusses the types of cryptoassets.

Central Bank Digital Currency (CBDC)

Central bank digital currency[113] is a digital form of fiat money (a currency established as money by government regulation, monetary authority or law). These differ from virtual or cryptocurrencies, which are not issued by a state and lack legal tender status declared by the government.

[110] Frankenfield, J.; "Lightening Network," Investopedia, 25 June 2019, *www.investopedia.com/terms/l/lightning-network.asp*
[111] Vogelsteller, F.; V. Buterin; "Ethereum Improvement Proposals," *https://eips.ethereum.org/EIPS/eip-20*
[112] ERC-721, *http://erc721.org*
[113] Wikipedia, "Central bank digital currency," *https://en.wikipedia.org/wiki/Central_bank_digital_currency*

ISACA. All Rights Reserved.

Cryptocurrency

Cryptocurrency is a digital representation of value that is not created or issued by a central bank or sovereign state, and can be used as a method of exchange. The most well-known example is bitcoin.

Security Token

A security token represents an equity, debt, derivative or other regulated security. These can be private or public securities. The security can be natively digital or can be a tokenized representation of a security that exists offchain. Most jurisdictions around the world consider these to be securities and subject to local securities laws.

These differ from Initial Coin Offerings (ICOs), which have been used previously to raise capital for startup companies, but have often been the subject of a wide variety of fraudulent schemes, and are subject to strict, sometimes criminal, penalties around the world.

Stablecoins

Stablecoins are cryptocurrencies that are designed to minimize the volatility of price of the stablecoin, relative to some stable asset or basket of assets. A stablecoin can be pegged to a cryptocurrency, fiat money or exchange-traded commodities (e.g., precious metals or industrial metals).[114]

Tokenized Asset

A tokenized asset represents another, asset such as real estate or art (tangible), or an intangible asset, such as intellectual property rights or royalties. Tokenized assets may be considered securities, depending on jurisdiction.

Utility Tokens

Utility tokens are nonsecurity tokens that are used in conjunction with blockchain-based applications and services. Usually, utility tokens offer holders the right to a service, product or platform, and can serve as incentives or rewards for certain behaviors. These tokens cannot be traded or exchanged like currency or securities.

5.8.4 Benefits of Tokenization

There are several benefits to tokenization:

- Tokenization increases the speed and reduces the costs of creating, issuing and exchanging assets.
- Reduced costs can decrease minimum investment amounts, allowing for an expanded pool of investors and potentially wider access to global markets for the unbanked and underbanked.
- Because blockchain technology and digital tokens allow for finer divisibility of assets (e.g., the smallest bitcoin unit is a Satoshi, or one-hundred-millionth of a bitcoin[115]), tokenized virtual and physical assets can potentially enable fractional ownership across a wide variety of previously inaccessible assets, such as fine art.
- Tokenization can potentially create vast amounts of new liquidity in markets for real-world assets or hard assets that are usually highly illiquid, such as real estate, precious gems and investment grade wine.
- Asset tokenization offers greater security and transparency for asset transactions. Because ownership is definitively recorded and tracked on a blockchain, and ownership rights are directly fixed within a token,

[114] Hayes, A.; "Stablecoin," Investopedia, *www.investopedia.com/terms/s/stablecoin.asp*
[115] Bitcoin Wikipedia, "Satoshi (unit)," *https://en.bitcoin.it/wiki/Satoshi_(unit)*

ISACA. All Rights Reserved.

blockchain and asset tokenization allow for more secure and traceable transactions, and enhance the ability to audit and value the assets across the globe.

Tokenization is a complex and rapidly evolving field and will continue to evolve global financial assets (including equities, bonds, derivatives and loans) and physical goods and commodities, such as gold, diamonds, sports collectibles and wine. Many believe that the full potential of tokenization to evolve markets and derive new business models has only begun to unfold.

5.8.5 Tokenization Example: Real Estate

The commercial real estate industry is notoriously inefficient, friction filled, error prone and expensive. With that, it is also important to note that global commercial real estate market represents a large portion of world economic asset and transaction activity. An MSCI report[116] from 2018 noted that the size of the professionally managed global real estate investment market rose to $8.5 trillion in 2017 (this number is surely higher now).

Much progress is now being made to use tokens and blockchain technology to grow and evolve the world of real estate transactions, given the importance of this global industry. One example for this market is using the Ethereum blockchain to create tokens that are programmed to include transaction history and ownership rights, along with smart contract powered rules to ensure asset issuance, distribution and transactions in a fully regulated and compliant manner.

For instance, there are rules to enforce not just the ownership of the asset, but also lockup periods, or the ability (or requirement) to transfer assets only to acceptable counterparties (bad on credit worthiness, for example). Specifically, early in 2020, the large Brazilian bank BTG Pactual publicized that it would use the Tezos[117] blockchain to issue a security token (via a Security Token Offering [STO][118]) that would be secured and backed by Brazilian real estate.[119] Other projects leveraging blockchain and real estate include land title registry, investor/tenant identity management and project financing.

5.9 Blockchain Validity and Audit Considerations

This section addresses the important though nascent topics of blockchain validity and audit considerations. An audit is a formal inspection and verification to check whether a standard or set of guidelines is being followed, records are accurate, or efficiency and effectiveness targets are being met. The purpose of an IT audit is to evaluate a system's internal control design and effectiveness, including but not limited to efficiency and security protocols, development processes, IT governance and oversight. Given this foundation, what are the important aspects of validating the use and functionality of blockchain, and what are the features of this technology that are subject to audit?

Audit considerations for blockchain mirror many of the usual audit items and concerns for standard information technology platforms. That said, the unique aspects of blockchain (both public and private), including the use of cryptography and peer-to-peer networks, the immutability of data, and the transparency of transactions, require some important audit considerations:

- Are there appropriate documents and specifications that explain the blockchain's architecture, design and build processes?[120]

[116] MSCI, 2017, *www.msci.com/www/research-paper/real-estate-market-size-2017/01032786497*
[117] Tezos, *https://tezos.com*
[118] Deloitte, "Security token offerings: The next phase of the financial markets evolution?," 2020, *www2.deloitte.com/content/dam/Deloitte/cn/Documents/audit/deloitte-cn-audit-security-token-offering-en-201009.pdf*
[119] Ledger Insights, June 2020, *www.ledgerinsights.com/brazilian-investment-bank-tokenizes-real-estate-blockchain/*
[120] Deloitte, "An internal auditor's guide to auditing blockchain," *https://www2.deloitte.com/us/en/pages/risk/articles/internal-auditing-guide-to-blockchain.html*

ISACA. All Rights Reserved.

- What testing has been done on the blockchain and/or smart contract platform? Have unit tests[121] and integration tests[122] been conducted? What were the results of these tests, and how are tests planned for execution in the future?
- Do the systems and applications designed to be a part of a blockchain implementation meet the overall business strategy and goals assigned to the system? What is a public or private blockchain mission, and how well aligned with that mission is the blockchain implementation?
- For an open source peer-to-peer platform, what are the controls necessary to ensure the viability and ongoing functioning of the blockchain?
- For public or private blockchain, who are the major support developers that maintain the code, allow for new code changes or forks, or help define new token standards? How deep and extensive is their collective history in these efforts?
- Has system access been segregated? Do execution duties on the platform change within roles over a period of time?
- How are security breaches managed and how is the blockchain system designed to manage attacks? (The DAO attack is a good example that warrants this concern.)[123]
- What is the consensus methodology used, and how relevant and applicable is it to the use case?
- What is the governance model associated with the specific blockchain? If it is a private blockchain, what parties are involved, how are they organized and what are their roles and functions? In addition, who determines the appropriate access for specific participants, and how is that enforced? For a public blockchain, how does the blockchain add new blocks of data, maintain the incentives for network management and ensure the veracity of transactions added to the ledger?
- What is the interoperable aspect of the specific blockchain, if any, and what are the governance rules associated with this interoperability?
- If using a private blockchain, what is the nature of trust within the system? Will data transfer be private within a group of well-known participants, and what group determines and validates those participants?[124]
- How does the blockchain and its commensurate smart contracts (if applicable) protect information assets and properly dispense information to authorized parties? For smart contract security audit, the following nonexhaustive list of questions must also be considered:[125]
  - Have security audits protocols and possible attacks been identified? Are race conditions or reentrancy attacks[126] known and have they been tested against and mitigated?
  - How is contract complexity being analyzed to minimize errors?
  - How is the smart contract coded to respond in the event of failures?
  - Are all tools and libraries updated, and what may occur if they are not?
  - For external calls of smart contracts specifically, how might these alter smart contract state and what is the impact of that alteration? How are errors in external calls handled?
  - Are any untrusted contracts or interactions appropriately flagged and isolated?

5.10 Conclusion

Blockchain technology and all of its ancillary solutions and services continue to evolve at an increasing pace. In addition, the adoption of blockchain technology across different industries around the world continues to accelerate.

[121] Software Testing Fundamentals, "Unit Testing," *http://softwaretestingfundamentals.com/unit-testing/*
[122] Software Testing Fundamentals, "Integration Testing," *http://softwaretestingfundamentals.com/integration-testing/*
[123] Siegal, D.; "Understanding the DAO Attack," Coindesk, 25 June 2016, *www.coindesk.com/understanding-dao-hack-journalists*
[124] Jade Research, "How to evaluate blockchain technology," *https://jaderesearch.org/how-to-evaluate-blockchain-technology*
[125] Blockgeeks, "How to Audit a Smart Contract," *https://blockgeeks.com/guides/audit-smart-contract/*
[126] *Ibid.*

ISACA. All Rights Reserved.

Although not an all-encompassing audit framework, given the important distinct features of blockchain technologies, cryptoassets, tokens and smart contracts, IT professionals are strongly encouraged to continue their education regarding this technology, because it continues to impact the world.

5.11 Chapter 5 Knowledge Check

REVIEW QUESTIONS

1. The Linux Foundation® and IBM® created an open source project for blockchain implementation known as:
   A. IBM Managed Blockchain
   B. R3
   C. Hyperledger®
   D. Ethereum®

2. __________ is a decentralized open source blockchain that enables the use of smart contracts.
   A. IBM Managed Blockchain
   B. R3
   C. Hyperledger
   D. Ethereum

3. One example of a native token is:
   A. A bitcoin
   B. Bitcoin Lightning
   C. Utility token
   D. ERC

4. ERC stands for:
   A. Error recording chain
   B. End run chain
   C. Ethereum request for comments
   D. Ethereum rally coin

5. A digital form of FIAT money is also known as:
   A. Central bank digital currency
   B. Cryptocurrency
   C. Asset tokenization
   D. Stablecoin

6. Blockchain systems do not need external auditing, as the technology is self-auditing.
   A. True
   B. False

ISACA. All Rights Reserved.

7. All of the following are important when auditing a blockchain system except:

   A. Interoperability capabilities
   B. Current testing being conducted
   C. Architecture, design and build process
   D. The value of a single cryptocurrency

Answers on page 77

ISACA. All Rights Reserved.

Chapter 5 ANSWER KEY

Review Questions

1. A. IBM Managed Blockchain
   B. R3
   C. Hyperledger. Refer to page 67.
   D. Ethereum

2. A. IBM Managed Blockchain
   B. R3
   C. Hyperledger
   D. Ethereum. Refer to page 66.

3. **A. A bitcoin. Refer to page 71.**
   B. Bitcoin Lightning
   C. Utility token
   D. ERC

4. A. Error recording chain
   B. End run chain
   C. Ethereum request for comments. Refer to page 71.
   D. Ethereum rally coin

5. **A. Central bank digital currency. Refer to page 71.**
   B. Cryptocurrency
   C. Asset tokenization
   D. Stablecoin

6. A. True
   B. False. Refer to pages 73-74.

ISACA. All Rights Reserved.

7. A. Interoperability capabilities
   B. Current testing being conducted
   C. Architecture, design and build process
   D. The value of a single cryptocurrency. Refer to page 74.

ISACA. All Rights Reserved.

Glossary

A

Altcoin—Have no formal definition but are widely considered to be alternative digital currencies; can also be all cryptocurrencies other than bitcoin

Application programming interface (API)—A set of routines, protocols and tools referred to as building blocks used in business application software development

Application-specific integrated circuits (ASIC)—A solid-state device designed to perform a single or small group of functions.

Atomic swaps—Peer-to-peer exchange of assets across separate blockchains triggered by predetermined rules, without the use of a third-party service, through the use of self-enforced smart contracts. Requires an exchange of assets on both sides or transaction will not occur

B

Blockchain—A distributed, protected journaling and ledger system. Use of blockchain technologies can enable anything from digital currency (e.g., Bitcoin) to any other value-bearing transaction

Blockchain explorers—Front end applications or user interfaces that allow a user to view individual records on a blockchain

Byzantine fault tolerant (BFT)—The property of a system that allows it to withstand failures and continue to function even if some of the nodes fail or act maliciously

C

Central bank digital currency (CBDC)—A digital form of fiat money

Central processing unit (CPU)—Computer hardware that houses the electronic circuits that control/direct all operations of the computer system.

Consensus mechanism—A fault-tolerant mechanism used in blockchain/distributed ledger systems to achieve the necessary agreement on data values or the state of the network among distributed processes or multiagent systems

Consortium blockchain—A subset of private blockchains that provides a unique blend of both public and private blockchain

Cryptoassets—Decentralized virtual currencies (and their underlying blockchain technology layers) that are meant to achieve something other than the exchange of value

Cryptocurrency—A digital asset designed and created to function as a unit of account and payment method within its particular ecosystem. Cryptocurrency transactions usually take place within a peer-to-peer network and use cryptography to secure transaction records.

Cryptography—The study of mathematical techniques related to aspects of information security, such as confidentiality, data integrity, entity authentication and data origin authentication

D

Decentralized autonomous organization (DAO)—A computer program on a blockchain that utilizes smart contracts to set organizational rules via decentralized means

DigiCash—An electronic money corporation and the private, secure digital money it delivers

E

Ethereum—An open source blockchain system enabling smart contracts and producing Ether as its native crypto currency

Ethereum request for comments (ERC)—Ethereum blockchain standards designed to enable Layer 2 tokens

F

Full node—A critical network device that supports and provides security for the blockchain and is capable of validating and relaying new blocks into the chain

G

Graphics processing unit—Special processing unit made to render high-quality images and video files

H

Hard fork—A change to blockchain software, such that any nodes validating according to the old software, will see all blocks produced subsequent to the new software as invalid. For blockchain nodes to work in alignment with the new software, each will be required to upgrade. If a group of nodes do not upgrade and perpetuate use of the old version of software, a permanent split in the blockchain can occur.

ISACA. All Rights Reserved.

Hash—A cryptographic hash function takes an input of an arbitrary length and produces an output (also known as a message digest) that is a standard-sized binary string. The output is unique to the input in such a way that even a minor change to the input results in a completely different output. Modern cryptographic hash functions are also resistant to collisions (situations in which different inputs produce identical output); a collision, while possible, is statistically improbable. Cryptographic hash functions are developed so that input cannot be determined readily from the output.

Hashed timelocks—A technical approach is a type of smart contract utilized in a cryptoasset transactions and is designed to remove counterparty risk, or the risk that the other party to a transaction cannot participate in the trade.

Hyperledger—An umbrella project started by the Linux Foundation, with participation by IBM, Intel and SAP, to build open source blockchains and related tools

I

Immutable—Unable to be modified after creation

Interoperability—The ability to exchange, access and make use of information across different systems and/or networks without the need for intermediaries. The capacity to transfer an asset between two or more networks or systems without changing the state of the asset.

L

Layer 2 tokens—A secondary coding that exists on top of the original blockchain coding structure that allows for the evolution of a decentralized blockchain to address limitations, i.e., scaling and smart contracts

Listening nodes—A publicly visible blockchain network device whose main function is to communicate and share data or information with any other node that connects with it

Litecoin—A peer-to-peer cryptocurrency and open-source software project

M

Masternodes—A blockchain network device that can process all the functions of a full node or miner but is also able to facilitate other processes

N

Native tokens—1. Created at the genesis block and usually used to reward the successful processing of a transaction or the creation of a blockchain; 2. the unit of account for a blockchain

Node—Point at which terminals are given access to a network.

Nonce—A limited or single-use, typically small value used as an initialization, seed or other special-purpose value.

P

Personally identifiable information (PII)—Any information that can be used to establish a link between the information and the natural person to whom such information relates, or that is or might be directly or indirectly linked to a natural person

Practical Byzantine fault tolerance (pBFT)—Consensus mechanism in which all nodes are ordered in sequence with one node being primary node or leader, and all others referred to as backup nodes. All nodes in pBFT systems communicate with one another with the goal being that all honest nodes will come to an agreement of the state of the system using a majority rule. Nodes communicate for two reasons: to prove that messages came from a specific peer node, and to confirm that that message was not modified during transmission; pBFT can be used for private and public blockchains and allows for instant transaction finality; however, such methodology requires a great number of messages between nodes, hence making a large blockchain network challenging.

Proof of elapsed time (PoET)—A consensus mechanism algorithm often used on permissioned blockchain networks to randomly decide the next block publisher

Proof of importance (PoI)—A variation of proof of stake that takes into consideration the role of validators and shareholders in the blockchain operation

Proof of stake (PoS)—Proof of stake is a type of consensus algorithm by which a cryptocurrency blockchain network aims to achieve distributed consensus. In PoS consensus, the creator of the next block of data is chosen via several combinations of random selection and wealth or age (i.e., the stake) within the blockchain; With PoS, miners can mine or validate block transactions based on amount of cryptocurrency a miner holds; was created as an alternative to PoW, which requires large amounts of energy; PoS gives mining power based on the percentage of cryptocurrency held by a miner; seen as less risky in terms of network attacks and security and used only for public blockchains.

Proof of Work (PoW)—PoW is conducted through miners (participants who keep the blockchain running by providing computing resources), who are competing to solve a cryptographic problem (hash puzzle). The PoW algorithm is used to confirm transactions and produce new blocks which are added to the chain. With PoW, miners compete against each other to complete transactions on the network and get rewarded. The computational work required to accomplish this is fairly (and usually increasingly) difficult for miners to perform, but easy for the network to verify. As difficulty increases over time, the amount of computational power, and, hence, energy consumption, grows. Bitcoin is the first widespread application use of PoW. PoW is applicable to public blockchains.

Pseudonymization—The processing of personal data in such a manner that the personal data can no longer be attributed to a specific data subject without the use of additional information, provided that such additional information is kept separately and is subject to technical and organizational measures to ensure that the personal data are not attributed to an identified or identifiable natural person

Public key encryption—A cryptographic system that uses two keys: one is a public key, which is known to everyone, and the second is a private or secret key, which is only known to the recipient of the message. See also Asymmetric Key.

S

Security token—Digital assets or tokens created to represent a quantity of a specified investment, including rights to ownership, payment of a specific sum under a contract, entitlement to future profits, etc.

Sidechain—A separate blockchain that links data entries or transactions to a primary blockchain, allowing operations both from and to the sidechain

Smart contract—Software (computer code) that automatically executes transactions and/or enforces agreements based on the fulfillment of the terms of the agreement by leveraging decentralized ledger technology that uses public validation to ensure correct and reliable performance according to agreed rules.

Soft fork—A software upgrade that is backward compatible with previous versions of the blockchain software. Thus, a soft fork does not require all blockchain nodes to upgrade to maintain functionality.

Stablecoins—A type of cryptocurrency that is tied to an outside, such as the US dollar, to stabilize its value

Supply chain management (SCM)—A concept that allows an enterprise to more effectively and efficiently manage the activities of design, manufacturing, distribution, service and recycling of products and service its customers.

T

Token—In security systems, a physical device that is used to authenticate a user, typically in addition to a username and password; in programming languages, a single element of the language

Topology—The physical layout of how computers are linked together

Scope Notes: Examples of topology include ring, star and bus.

Transparency—Refers to an enterprise's openness about its activities and is based on the following concepts:

- How the mechanism functions is clear to those who are affected by or want to challenge governance decisions
- A common vocabulary has been established
- Relevant information is readily available

Scope Notes: Transparency and stakeholder trust are directly related; the more transparency in the governance process, the more confidence in the governance.

U

Utility token—Digital assets or tokens created and utilized to finance creation of a network by providing its buyers with a pledge of being able to use some of the network ecosystem or products; do not give any legal or economic right of ownership over the developer nor any part of the ecosystem

V

Validation—Establishing documented evidence that provides a high degree of assurance that a specific process will consistently produce a product meeting its predetermined specifications and quality attributes

ISACA. All Rights Reserved.

Virtual currency—Digital representations of value, not created or issued by a central bank or sovereign state, which can be used as a method of exchange

W

Wallet—An application or other service that gives holders of cryptocurrency the ability to store and retrieve their digital assets. Such wallets come in many forms, including hot wallets (any wallet application or service connected to the Internet), or cold wallets (or cold storage, which are often hardware devices that can be disconnected from the Internet or other electronic services).

9781604207576